Workbook for
Fundamentals of Nursing

Fourth Edition

Barbara Kozier, RN, MN

Glenora Erb, RN, BSN

Lina K. Sims, MSN, RNCS
Certified Clinical Specialist, Adult Psychiatric and
Mental Health Nursing
Assistant Professor, Saint Joseph College of Nursing

Addison-Wesley Nursing
A Division of The Benjamin/Cummings Publishing Company, Inc.
Redwood City, California • Menlo Park, California
Reading, Massachusetts • New York • Don Mills, Ontario • Wokingham
Amsterdam • Bonn • Sydney • Singapore • Tokyo • Madrid • San Juan

To Marc, Robyn, and Renee, whose love, faith
and support make all things possible.

The author and publishers have exerted every effort to ensure that all content, related nursing interventions and drug selection and dosages set forth in this text are in accord with current recommendations and practice at the time of publication. However, in view of ongoing research, changes in government regulations, and the constant flow of information relating to the practice of nursing, drug therapy and drug reactions, the reader is urged to check the package insert for each drug for any change in indications of dosage and for added warnings and precautions. This is particularly important where the recommended agent is new and/or infrequently employed. In addition, all persons included in case studies in this book are fictional and not intended to represent any persons living or dead.

ISBN 0-201-51661-6

12345678910-AL-95 94 93 92 91

Addison-Wesley Nursing
A Division of The Benjamin/Cummings Publishing Company, Inc.
390 Bridge Parkway
Redwood City, California 94065

PREFACE

As students pursuing an education in professional nursing, you are faced with myriad concepts, principles, and information that must be assimilated within a short period of time. This *Study Guide for Fundamentals of Nursing* was written to help you to learn and apply the content presented in *Fundamentals of Nursing, Fourth Edition,* by Kozier, Erb, and Olivieri. Using this study guide will enable you to grasp the material as efficiently and as effectively as possible. Each chapter in the study guide corresponds with a chapter of *Fundamentals of Nursing,* and includes the following sections:

- **Introduction:** a brief summary of the major areas covered in the text chapter.

- **Learning objectives:** a list of competencies that you will be able to accomplish after reading the textbook and completing the study guide chapter. Each exercise in the chapter has been developed to help you attain one or more of the learning objectives.

- **Exercises:** a variety of stimulating and thought-provoking exercises that will help you learn the textbook material. Each study guide chapter is broken into sections that correlate with the major parts of the text chapter: these sections are indicated by an open-book icon.

- **Self-assessment questions:** a section of multiple choice questions is located at the end of each chapter for you to test your knowledge.

- **Learning activities:** individual, classroom, and clinical activities are recommended for those of you who enjoy additional challenges to learning.

An answer section for immediate feedback is located in the appendix.

I hope that you will enjoy using this study guide and that it will help to prepare you for managing the vast amount of information needed to provide quality nursing care.

ACKNOWLEDGEMENTS

This book could not have been completed without the assistance of Devra Lerman, whose unlimited patience, support, and encouragement walked me through the steps of this project; Patti Cleary, whose editorial expertise and foresight initiated this edition of the study guide; Marc Sims, whose technical computer support made this work possible; and all my students past and present who make being a nurse educator new and exciting each day!

Lina K. Sims

TABLE OF CONTENTS

INTRODUCTION TO NURSING

Chapter 1 introduces the student to the profession of nursing. After reviewing the chapter, participating in classroom discussion and activities, and completing the learning guide exercises the student will be able to:

- Identify the essential aspects of nursing.
- Explain professional growth within nursing.
- Identify the functions that Styles cites as necessary for the preservation and development of a profession.
- Identify the critical attributes of professionalism in nursing as identified by Miller.
- Identify essential concepts in the historical development of nursing.
- Explain the significance of nurse practice acts.
- Describe the settings for nursing practice.
- Describe the importance of standards of nursing practice.
- Explain career mobility and expanded roles.
- Describe the various educational programs in nursing.
- Explain the functions of the national nurses' associations.

NURSING: AN EMERGING DEFINITION

Florence Nightingale was one of the first nursing theorists to define nursing. She and a host of other distinguished nurses over the past 100 years have contributed to the definition of nursing as it is known today. After reviewing nursing definitions in Chapter 1, list and briefly describe at least five themes common to the various definitions of nursing in use today:

1. ______________________________

2. ______________________________

3. ______________________________

4. ______________________________

5. ______________________________

NURSING: AN EVOLVING PROFESSION

Identify each of the following persons or groups that were significant in the development of the nursing profession:

6. ____________________ proposed a study of nursing education that led to the accreditation process for schools of nursing.

7. ____________________ wrote a nursing text in the late 1800s that became a standard for nursing schools in America.

8. ____________________ established a school of nursing in London which is credited with providing the first planned educational program for nurses.

9. ____________________ America's first trained Afro-American nurse.

10. ____________________ Recommended that schools of nursing be independent of hospitals and on a college level.

Check the following persons who are members of a profession.

11. _______ Veterinarian
12. _______ Nurse
13. _______ Medical Doctor
14. _______ Secretary
15. _______ Lawyer
16. _______ Salesman

Complete the tables below describing Miller's three critical attributes of professionalism in nursing and five behaviors displayed by persons who practice professional nursing.

CRITICAL ATTRIBUTES OF PROFESSIONALISM
17.
18.
19.

PROFESSIONAL BEHAVIORS
20.
21.
22.
23.
24.

In Chapter 1 Styles delineates five functions that nursing organizations must perform for the preservation and development of the profession. Describe the five functions in the space provided:

25. ______________________________

____

____

____

26. ______________________________

____

____

____

27. ______________________________

____

____

____

28. ______________________________

____

____

____

29. ______________________________

____

____

____

NURSING PRACTICE

The diagram below depicts the four areas of nursing practice related to health. Fill in each quadrant with one of the four areas that focuses on health.

30.	32.
31.	33.

In the past, nurses practiced primarily in hospital settings. Today, nurses practice in a variety of diverse settings. In the space below list at least five areas nurses practice in outside of the hospital setting:

34. ____________________

35. ____________________

36. ____________________

37. ____________________

38. ____________________

In the spaces provided match each term in column I with the appropriate description in column II:

39. _______ Nurse specialist
40. _______ Primary nursing
41. _______ Standards of practice
42. _______ Client
43. _______ Nurse practice act
44. _______ Case management
45. _______ Case method
46. _______ Patient
47. _______ Nurse practitioner
48. _______ Functional method
49. _______ Consumer
50. _______ Team nursing
51. _______ Nurse clinician

A. Individual who uses a service or commodity
B. Comes from a latin word meaning to suffer or to bear
C. Person who engages the advice or service of another who is qualified to provide this service
D. Total nursing care
E. Focuses on nursing jobs that need to be completed
F. Provides bedside or direct care in a specialty area
G. One nurse is responsible for total care of clients 24 hours a day
H. Requires a nurse with a BS or MS to implement high level professional care
I. A formalized contract between society and the nursing profession
J. Master-prepared nurse who has advanced knowledge and skill in a particular area of nursing
K. Delivery of individualized nursing care by a team led by a professional nurse
L. Nurse with advanced degree who provides specialized care
M. Criteria against which clients, nurses, and employers can evaluate nursing care

EDUCATION FOR NURSES

True and False

52. _______ state laws in Canada and the United States recognize two types of nurse: the licensed practical nurse and the registered nurse.

53. _______ The terms licensed practical nurse (LPN) and licensed vocational nurse (LVN) are terms used for the same level of nursing.

54. _______ Both Canada and the United States provide nursing education in three types of programs: diploma, associate degree and baccalaureate degree programs.

55. _______ Most diploma programs are offered in junior colleges as well as colleges and universities.

56. _______ Continuing education programs allow graduate nurses to obtain advanced degrees.

57. _______ Inservice education is usually administered by an employer to upgrade knowledge and skills of employees.

NURSING ORGANIZATIONS

In the space provided, identify the organization represented by the descriptions listed below:

58. ______________________ was established in 1970 by the ANA to recognize nurses who have made significant contributions to the profession.

59. ______________________ is a preprofessional organization for student nurses.

60. ______________________ is a federation of nurses' organizations from different countries.

61. ______________________ is an agency of the United Nations whose primary purpose is to bring high level health care to people of the world.

62. ______________________ is an international honor society in nursing.

63. ______________________ includes both nurses and non-nurses in its membership, who join together to foster the development and improvement of nursing services and nursing education.

64. ______________________ has developed national standards and a code of ethics for Canadian nurses.

65. ______________________ is the national professional organization for nursing in the the United States.

SELF ASSESSMENT QUESTIONS

66. The first journal to publish findings of nursing studies was

    A. *Nursing Research.*
    B. *Advances in Nursing Science.*
    C. *American Journal of Nursing Education.*
    D. *Image: Journal of Nursing.*

67. Autonomy in a profession means members

    A. develop their own curricula.
    B. determine who is allowed into the profession.
    C. regulate and set standards for themselves.
    D. set fees for the public.

68. Health promotion means

    A. helping people develop resources to maintain their well-being.
    B. helping clients maintain their health status.
    C. helping people improve their health after an illness.
    D. comforting and caring for people who are dying.

69. The American Nurses' Association

    A. is the national professional organization for nurses in the United States.
    B. includes members from all South and North American countries.
    C. evaluates schools of nursing for excellence.
    D. is an official agency of the United Nations.

70. "Client" is the preferred term for a recipient of nursing care because the term implies that the recipient is a

    A. passive participant in the nursing care process.
    B. collaborator in the nursing care process.
    C. receiver of care when sick or dying.
    D. negotiates fees with care givers.

ADDITIONAL LEARNING ACTIVITIES

1. Write your own personal definition and philosophy of nursing.

2. Interview a graduate nurse from a diploma, an associate degree and a baccalaureate program. How do these nurses`differ in their practice of nursing? How are they similar?

Crossword Puzzle: Chapter 1

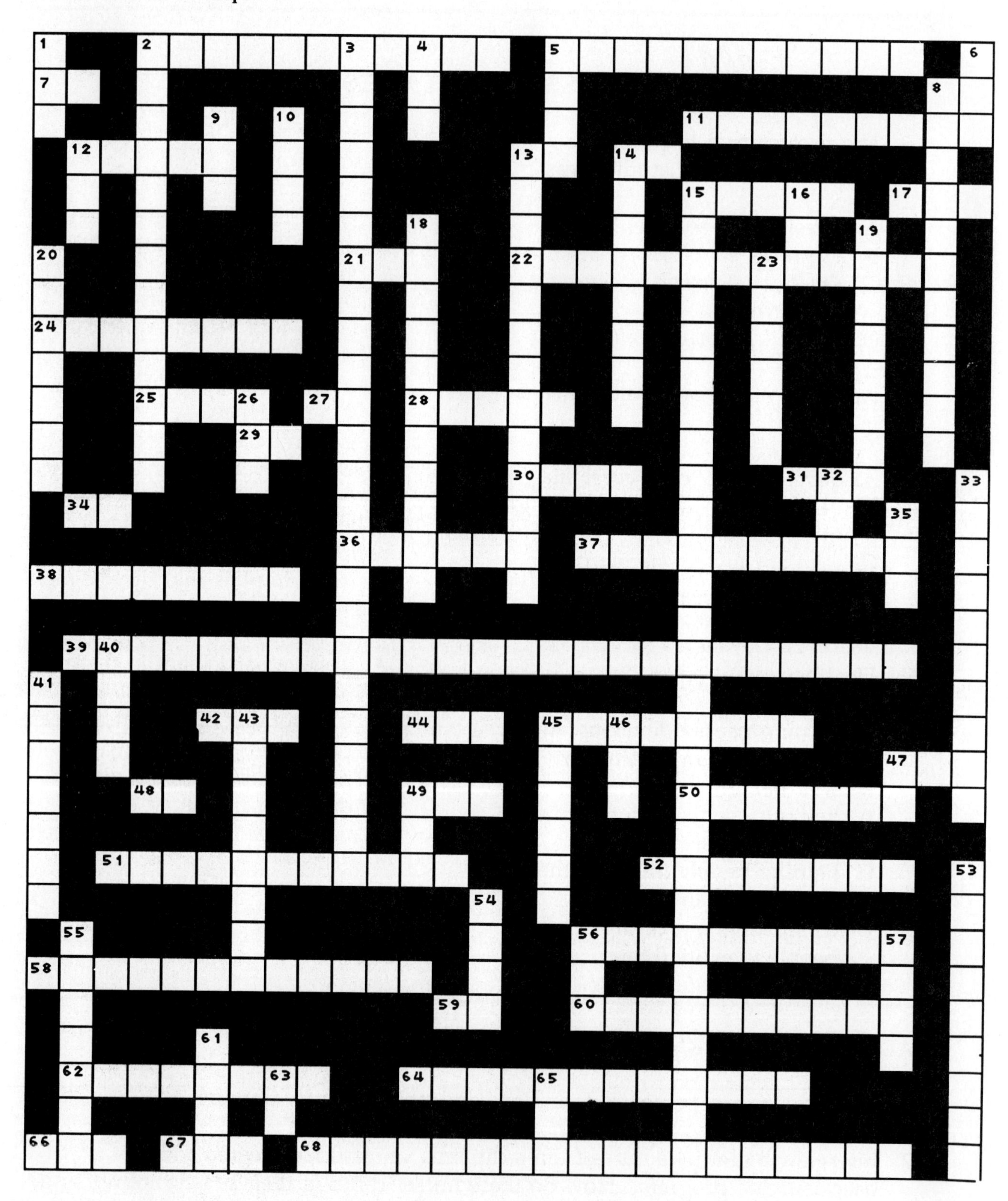

Across

2. She and 64 across cared for the New York poor
5. One of the four areas of nursing practice related to health
7. Doctor of nursing (abbr)
8. Same as 63 down
11. Provides care in a specialty area
12. ________ of ethics
13.* Northeast (abbr)
14.* Intramuscular (abbr)
15. _______:*The Journal of Nursing Scholarship*
17.* Visual perception organ
21. American Nurses' Association (abbr)
22. International honor society in nursing
24. America's first trained black nurse
25. Organized first Johns Hopkins school of nursing
27. Master of arts
28. Licensure examination
29. Member of the NSNA
30. Registered nurse clinical specialist (abbr)
31. Doctor of philosophy
34. Nurse practitioner (abbr)
36.* Indwelling catheters
37. One of the first modern nurses to define nursing
39. Professional US nursing organization
42. Licensed vocational nurse (abbr)
44. An agency of the United Nations (abbr)
45. An individual who uses a service or commodity
47. Health maintenance organization (abbr)
48. Same as 7 across
49. *Advances in Nursing Science* (abbr)
50. A collaborator in nursing care received
51. Along with 52 across, the first nursing theorist
52. The lady with a lamp
56. A calling that requires special knowledge, skill, and preparation
58. Established the FNS after World War I
59.* Doctor of medicine (abbr)
60. An activity in which one engages
62. The ability of a professional group to regulate itself
66. Same as 17 across
67. Nurse practice ___: a contract between society and nursing
68. The first nursing order established in the Middle Ages

Down

1. Frontier Nursing Service (abbr)
2. America's first trained nurse
3. Accredits schools of nursing
4. American Academy of Nursing (abbr)
5. _______ management
6. Master ofscience in nursing
8. Lengthened nurses' training to three years
9. Same as 6 down
10. Fellow of the American Academy of Nursing
12. Canadian Nurses' Association (abbr)
13. Seeks to define basic principles of nursing practice
14. 25 across's first initial, middle name
15. A federation of national nurses' associations
16.* To obtain
18. The first president of the CNA
19. Measures nursing competence
20. 14 across without the "I"
23. Concerned with the worth and dignity of man
26. Same as 65 down
32.* Possesive male gender
33. Another name for "total care"
35. Doctor of nursing science
40. 58 across's first name
41. Hospital-based program
43. Work that a person regularly performs
45. _________ mobility
46. National League For Nursing (abbr)
47.* Bedtime
49.* Total
53. Health _________
54. Nursing meets the ______s of the individual
55. One nurse is responsible for total care
56. Health ___motion
57. National Student Nurses' Association (abbr)
61. Doctor of nursing science
63. Master of science (abbr)
65. Bachelor of science degree in nursing (abbr)

* Terms not found in Chapter 1

2 SOCIALIZATION AND ROLES OF THE NURSE

Chapter 2 focuses on the socialization experience the beginning nurse encounters when entering the profession. Values that are critical to the practice of professional nursing, various types of socialization processes, as well as specific roles that the nurse assumes are explored. After reviewing this chapter the student will have a more realistic picture of the profession of nursing and will be able to:

- Differentiate primary, secondary, and anticipatory socialization.
- Describe the process of professional socialization.
- Identify four critical values of professional nursing.
- Compare the socialization models of Simpson, Davis, Hinshaw, and Dreyfus.
- Compare Dalton's and Kramer's models of career development.
- Discuss essential aspects of the nurse's role.

TYPES AND CHARACTERISTICS

Explain the meaning of the following terms in your own words:

1. Socialization ______________________________

____

2. Primary socialization ______________________________

____

3. Reciprocal socialization ______________________________

____

4. Secondary or adult socialization ______________________________

____

5. Anticipatory socialization ______________________________

____

6. Resocialization ____

____

PROFESSIONAL SOCIALIZATION

Describe the various ways that the "Process of Professional Socialization," is defined in Chapter 2 by:

7. Watson ____

____

8. Hinshaw ____

____

9. Styles ____

____

In the diagram below, identify Watson's four critical values of professional nursing.

10.

11.

12.

13.

Fill in the phases for each model of socialization into professional roles in the appropriate boxes.

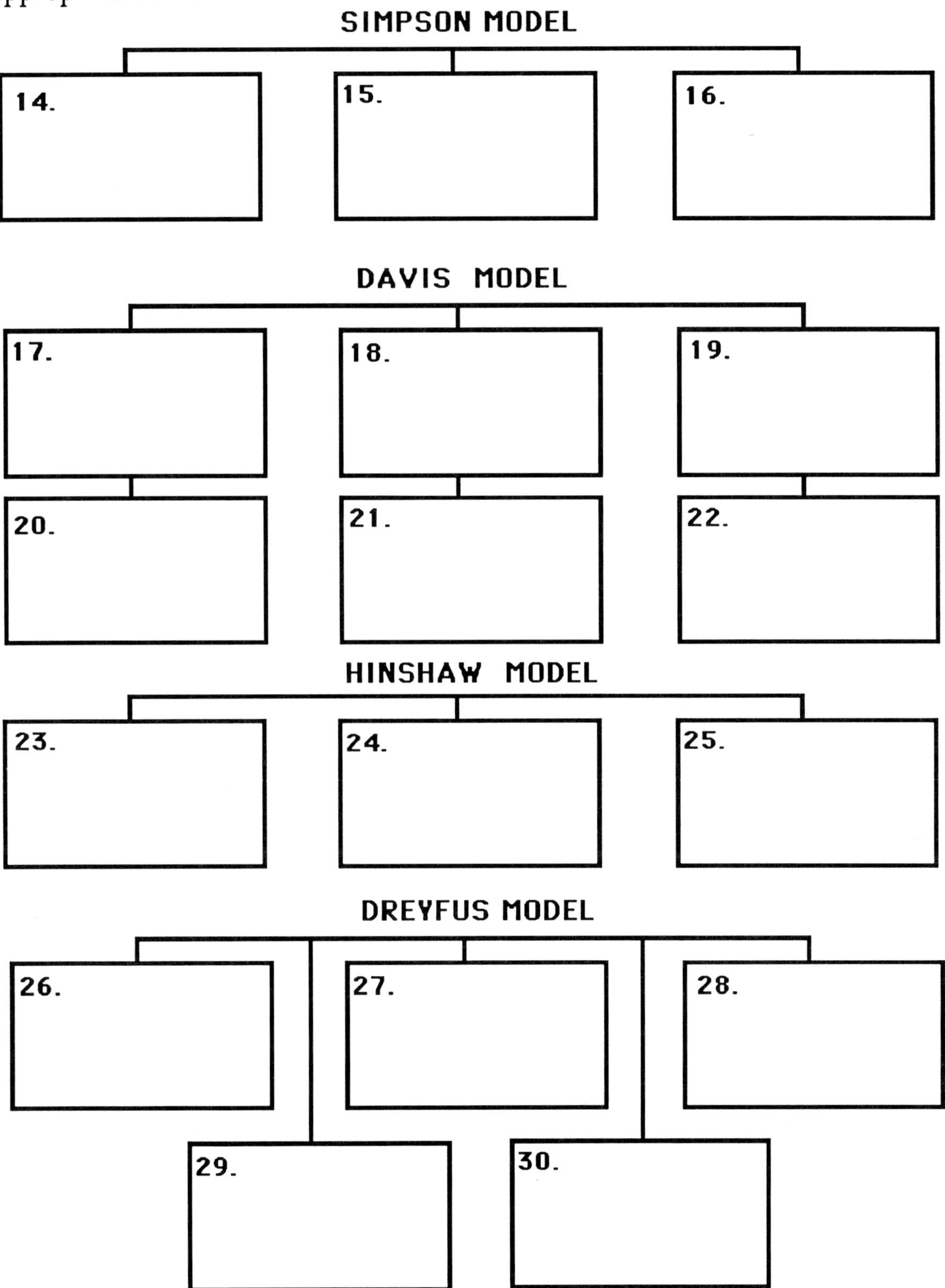

Dalton and Kramer each proposed a four stage model of career development cited in Chapter 2. In the space below write a brief comparison of each stage of the two models:

STAGE	DESCRIPTION
31. Stage I	
32. Stage 2	
33. Stage 3	
34. Stage 4	

ROLES OF THE PROFESSIONAL NURSE

In the spaces provided match each term in column I with the appropriate description in column II.

35. ______ Counselor
36. ______ Carer
37. ______ Teacher
38. ______ Advocate
39. ______ Communicator

A. Shows concern for the client

B. Listens, responds, and develops relationships with clients

C. Pleads the cause of another

D. Provides emotional, intellectual, and psychologic support

E. Achieves behavior changes in a client

SELF ASSESSMENT QUESTIONS

40. Socialization is a process where people

    A. learn to become members of society.
    B. develop a language from two cultures.
    C. join an ethnic group.
    D. develop a set of governmental rules.

41. Primary socialization occurs from

    A. birth to school age.
    B. birth to adolescence.
    C. birth to young adulthood.
    D. birth to marriage.

42. Nancy, a four year old, likes to play house with her next door neighbor, Tommy. Nancy plays mother and Tommy plays father. This is an example of

    A. primary socialization.
    B. anticipatory socialization.
    C. resocialization.
    D. reciprocal socialization.

43. Which of the following descriptions *best* describes the role of the client advocate? A clinical advocate
    A. acts as a protector when the client is faced with potential professional incompetence.
    B. provides legal counsel in cases of malpractice.
    C. ensures that client's needs are met and rights are protected.
    D. interprets medical terminology so that client is well informed.

44. Georgia Mitchell, RN, an administrator of a large outpatient surgical clinic, lays off five surgical technicians because of budgetary constraints. Several of the nurse managers are angry because they had no input into the decision. Georgia says, "I'm sorry that you are upset but this was an administrative decision." Georgia's leadership style is

    A. autocratic.
    B. democratic.
    C. laissez-faire.
    D. selective.

ADDITIONAL LEARNING ACTIVITIES

1. Interview a nurse who functions in the role of a client advocate. Determine how the advocacy role is incorporated into the professional nursing role by the nurse.

2. Interview a mother and father with young children. Determine how they teach their children how to get along with others. How do the children learn social and religious values?

Crossword Puzzle: Chapter 2

Across

1. A focus of nursing
4. Type of leadership
7.* Medical ethics (abbr)
8. An _______ protects clients' interests
12.*Organs of perception
14. A nurse's role
15.*Client sign or symptom
16. Registered nurse (abbr)
17.*Intensive _ _ (abbr)
18. What a nurse manager does
19.*American Nurses' Association (abbr)
20. Leadership style
22. _______ mary
23. What a nurse manager does
25.*Canadian Nurses' Association (abbr)
28. Same as 22 across
29. __________ agent
31. What 1 down does
33. Initial inno____________
34. What a nurse manager has
36. Learned in acculturation
39. Dreyfus 5th stage of socialization
41. Implements nursing studies
45. A type of socialization
46. A teacher is one type of nursing _______________

Down

1. The communicator/ role _______________
2.* Morning hour
3. See 18 across
4. Professional ___________
5.* 32 = 1 quart
6. Another leadership style
9. See 36 across
10. ______________ styles
11. __________ socialization
13.*Associate nurse (abbr)
21. Change _______________
24.*Emergency room (abbr)
27. Socialization controls this
29.*Clinical specialist (abbr)
30. Mature person
32. See 24 across
35. First
37.*United Nations (abbr)
38.*Hospital sign
40. Anticipatory socialization (abbr)
42.*See 12 across
43.*What female nurses wear
44. Democratic leadership (abbr)

3 CHANGING NURSING PRACTICE

Chapter 3 deals with the concept of "change" as it applies to the profession of nursing in today's world. In order to stay current with advanced nursing knowledge, expanding medical technology, and an ever-changing health care delivery system, the nurse must be able to recognize the need for, as well as be willing to be, an agent of change. After completing Chapter 3, the student will be able to:

- Describe Bennis, Benne, and Chin's strategies for change.
- Describe the change process.
- Identify the factors influencing nursing practice.
- Explain how nursing education affects nursing practice.
- Describe the role of nursing research as it influences current and future nursing practice.
- Describe the role of the nurse in protecting the rights of clients.
- Identify the attitudes and behaviors necessary to overcome powerlessness.
- Explain the various sources of power.
- Describe why nurses should be politically active.
- Describe the guiding principles for political action.

CHANGE

List the eightsteps of the Change Process outlined in Chapter 3:

1. ______________________ 5. ______________________

2. ______________________ 6. ______________________

3. ______________________ 7. ______________________

4. ______________________ 8. ______________________

Bennis, Benne, and Chin describe three catgories of strategies to implement change: power-coercive, empirical-rational, and normative-reeducational. Identify which strategy is used for each of the following situations:

9. ______________________ A diabetic client has trouble following his prescribed diet. He tells the nurse, "Its too much trouble measuring and weighing everything I eat." The nurse teaches the client how to estimate amounts of food, reducing the time he had previously spent preparing meals.

10. ______________________ A student nurse states, "I hate taking care of elderly people. They're so slow!" An older nurse who the student admires takes the student to a geriatric rehabilitation unit and explains the positive aspects of caring for elderly clients.

11. ______________________ A RN team leader decides that no members of the team may take a morning break until all clients have had their AM care.

FACTORS INFLUENCING NURSING PRACTICE

Briefly describe how each of the following factors have influenced the practice of nursing:

12. Economics ____

____

13. The nursing shortage ____

____

14. Consumer demands ____

____

15. Family structure ____

____

16. Science and technology ____________________

____

17. Legislation ____________________

____

18. Demography ____________________

____

19. The women's movement ____________________

____

20. Collective bargaining ____________________

____

21. Nursing associations ____________________

____

NURSING EDUCATION

True and False.

22. ________ "Grandfathering" is a term used to describe a method of protecting nurses already licensed to practice from new credentialing legislation.

23. ________ Most practicing graduates of diploma and associate degree programs oppose the "grandfather" addition to state nursing legislation.

24. ________ Since 1985 the ANA has proposed two levels of nursing practice: registered nurse and associate nurse.

25. ________ Certification for clinical specialties in nursing must be attained through the American Nurses' Association.

26. ________ Basic nursing education programs are designed to prepare nurses as generalists as well as specialists in various aspects of nursing.

27. ________ It is the responsibility of the ANA to legislate changes in nursing education in various states.

28. ________ A nursing specialty is a defined area of clinical practice that has a narrow in-depth focus.

NURSING RESEARCH

List at least five resources for reviewing current nursing research in the literature:

29. ______________________________

30. ______________________________

31. ______________________________

32. ______________________________

33. ______________________________

Case Study:

Martha Jones, an eighty-seven-year-old retired school teacher, was admitted to Cunningham County Medical Center with severe ulcerations on her lower extremities. She has lived in a small federally subsidized apartment since her retirement nearly thirty years ago. She has no family since she never married and was an only child. Her next door neighbor, Madge Johnson, accompanied Martha to the hospital. Madge told the admitting nurse that Martha has been very confused recently and forgets to eat unless she is reminded. Several days after Martha's admission, Mary Adams Martha's primary nurse, asked the client if she would be willing to participate in a study the medical center was conducting. She explained to Martha that the purpose of the study was to determine whether a new treatment recently developed at the medical center would be more effective in treating leg ulcers. The nurse asked Martha to sign a form which gave the medical center permission to include her in the study. The following questions relate to Ms. Jones.

What information should Martha receive before she signs the consent form?

34. ______________________________

Martha says to the nurse, "I guess I'll get better care if I sign this form." How should the nurse respond?

35. ______________________________

List at least four rights Martha is entitled to as a subject in this research study:

36. ______________ 38. ______________

37. ______________ 39. ______________

Does Mary Adams as a primary nurse have any special reponsibility to a client who is in a research study?

40. ______________________________

How might this particular research influence nursing practice in the future?

41. ______________________________

____

____

____

The first step in the research process is to identify the problem. the following figure depicts the remaining nine steps. Fill in the blanks where indicated:

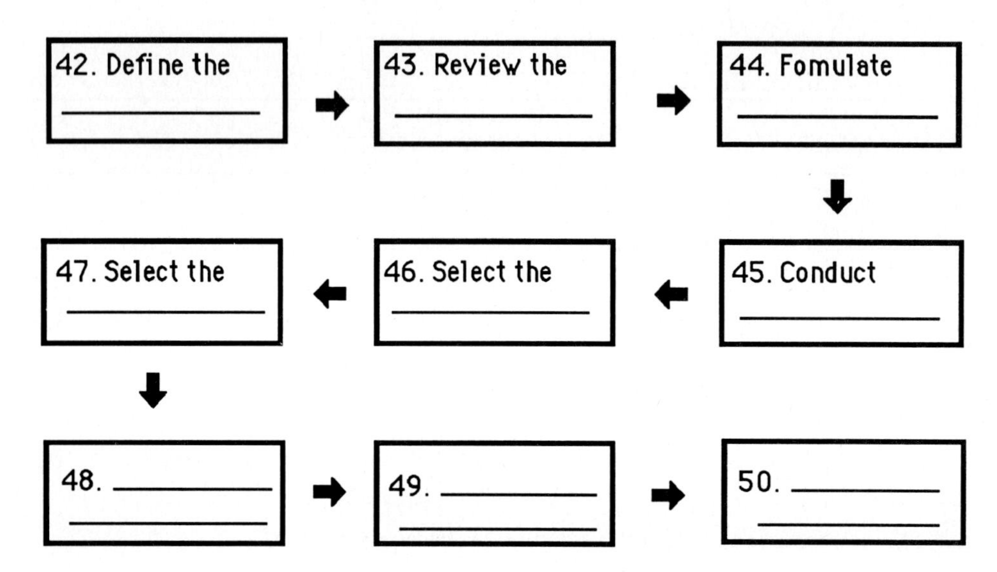

TRENDS IN NURSING

One of the major trends in nursing today is the increasing importance of the computer in enhancing nursing care delivery. List at least four ways a computer is used to improve nursing practice.

51. ____________________ 53. ____________________

52. ____________________ 54. ____________________

EFFECTING PROFESSIONAL CHANGE

The scramblegram below contains terms related to change theory discussed in Chapter 3. Identify each term in the definition section below, then circle the term in the scramblegram. The words may appear horizontally, vertically, diagonally, backwardsor forward.

P	R	O	A	C	T	I	V	E	G	I
B	A	C	T	I	O	N	C	O	D	N
R	E	F	O	U	R	V	V	W	O	F
E	X	L	M	E	N	E	O	I	P	L
F	Q	R	W	S	R	T	T	U	V	U
E	G	O	H	N	I	C	N	A	J	E
R	P	K	M	L	E	M	I	N	L	N
E	M	C	N	N	O	F	G	V	H	C
N	N	A	N	A	J	K	M	N	E	E
T	S	O	T	U	T	R	E	P	X	E
X	C	O	M	M	U	N	I	T	Y	V

55. ______________________ The capacity to modify the conduct of others in a desired manner.

56. ______________________ A type of power which arises from the perception of one's ability to threaten, harm, or punish others.

57. ______________________ A charismatic or personal type of power.

58. ______________________ Power derived from having important contacts or relationships with others.

59. ______________________ May Include the local neighborhood, the corporate world, and the nation.

60. ______________________ The bringing about of change.

61. ______________________ A forum for political action.

62. ______________________ American Nurses Association (abbr)

63. ______________________ Canadian Nurses Association (abbr)

64. ______________________ National League for Nursing (abbr)

65. ______________________ Power derived from one's expertise, talents, and skills.

66. ______________________ the result of the proper use of power

Complete the following sentence: Nurses should be politically active because:

67. ____

____

____

____

____

Jerri Martinson, RN, is a the unit manager in the surgical intensive care unit in a large metropolitan hospital. In the past year Jerri has lost four excellent nurses to another hospital in the community that pays higher wages and provides better fringe benefits. Jerri is concerned that she will continue to loose valuable personnel unless some changes in the employee wage and benefit package are made. Jerri is frustrated because discussions with administration concerning this problem have led nowhere. In the space below list and describe at least three attitudes and/or skills can Jerri use to bring about a change in this situation.

68. ____

____

69. ____

____

70. ____

____

SELF ASSESSMENT QUESTIONS

71. Donna Smith works in a free clinic that serves hispanic clients. Donna is often frustrated because her clients seek medical help from the local curandero (Hispanic folk healer) before coming to the clinic. Which of the following strategies would be most effective in changing the help-seeking behavior of Donna's clients?

    A. Establish a policy that clients who see the curandero may not be treated at the clinic.
    B. Enlist the curandero's help in convincing the clients that they should seek medical assistance sooner.
    C. Develop an educational program designed to teach the clients the benefits of medical science.
    D. Plan a private meeting with each client who regularly see s the curandero.

72. Developing an educational program that teaches Donna's clients the value of medical science is an example of ___________ change strategy

    A. power-coercive
    B. empiric-rational
    C. normative-reeducative
    D. none of the above

73. What impact have the recent changes in the health-care delivery system in the United State had on the practice of nursing?

    A. The clients nurses care for in hospitals are chronically ill.
    B. Less nursing care is being provided in the client's home.
    C. Nursing care is becoming less expensive.
    D. Health promotion is becoming increasingly important.

74. Martha Jone's (see case study and questions 34 - 50) rights may have been violated if she gave consent to be included in the research study because Martha

    A is too old to be involved in a research study.
    B. is confused at times.
    C. is critically ill.
    D. has no family.

75. The major role of the nurse caring for a client like Martha Jone's who is a subject in a research study is to

    A. collect valid data.
    B. protect the client's rights.
    C. report findings.
    D. communicate discrepancies.

ADDITIONAL LEARNING ACTIVITIES

1. Organize a group of students to travel to your state capitol to talk to state legislators about impending legislation concerning the profession of nursing or health-care delivery.

2. Write letters to your local politicians concerning health care or professional nursing issues in your area.

3. Visit a session of your county representatives. Observe the process used to bring about new legislation.

4. Identify a situation in your life that you would like to change. What attitudes and skills would you have to employ to bring about this change?

THEORIES / CONCEPTUAL FRAMEWORKS

From the earliest of times nurses used the intuitive knowledge passed down from generation to generation as a basis for nursing practice. In the mid 19th century, Florence Nightingale began to document the body of knowledge that was specific to nursing. Since that time nurses continue to conceptualize, research and document the science of nursing. Chapter 4 introduces the learner to this process. After reviewing this chapter the learner will be able to:

- Explain the purposes of nursing theories.
- Identify the concepts that form the metaparadigm of nursing.
- Differentiate a theory from a conceptual framework.
- Identify three essential elements of a theory.
- Identify three essential components and seven major units of a conceptual model of nursing.
- Describe the relationship of the nursing process to conceptual models of nursing.
- Describe the relationship of nursing theory to nursing research.
- Compare selected conceptual models for nursing.
- Explain the relationship of holism to nursing.
- Identify selected characteristics of basic human needs and factors influencing priority of needs.
- Discuss various constructs of caring.
- Describe general systems theory.
- Describe various approaches of problem solving.
- Identify three phases of the decision-making process.
- Explain the relationship of perception theory to the nursing process.

THEORIES AND CONCEPTUAL FRAMEWORKS

List the steps of the nursing process.

1. ______________________ 4. ______________________

2. ______________________ 5. ______________________

3. ______________________

Understanding the following terms is essential for comprehending nursing theories and conceptual frameworks. In the spaces provided match each term in column I with the appropriate description in column II.

6. _______ Concept
7. _______ Metaparadigm of nursing
8. _______ Model
9. _______ Framework
10. _______ Conceptual framework
11. _______ Theory
12. _______ Assumption
13. _______ Value system

A. Derived from scientific theory or practice or both

B. A basic structure supporting anything

C. A pattern of something to be made

D. Beliefs underlying a profession

E. An abstract idea or mental image , phenomena or reality

F. Integration of a set of concepts and statements into a meaningful configuation

G. Person, health, environment, and nursing action are components

H. Purpose is to generate knowledge

I. Provides a guide for nursing practice, education and research

Complete the following statements by placing the appropriate term(s) in the space provided:

14. A ____________________ is an abstract, idea or mental image of reality.
15. Conceptual models for nursing are abstractions that are made real by the use of the ______________
16. The primary purpose of nursing theory is to generate ______________
17. In the ____________method, a theory is devised and then research is conducted.

18. In the __________________ method, research is conducted first and the findings are used to develop a theory.

19. The major distinction between a theory and a conceptual model is the __________________

20. Nursing theory and nursing arch are closely related because the purpose of nursing theory is to generate scientific knowledge and scientific knowledge is derived from ____________________

Briefy describe the seven major units of a conceptual model of nursing listed below.

21. Goal of nursing____

____

22. Client ____

____

23. Role of the nurse ____

____

24. Source of difficulty ____

____

25. Intervention focus ____

____

26. Modes of intervention____

____

27. Consequences ____

____

THEORETICAL VIEWS OF HUMAN BEINGS

Complete the following paragraph on the systems and holistic theories by filling in the blanks.

Two very different and diverse ways of looking at the concept of man are through *systems* and *holistic* theories. Roy states that the human being is an 28__________ constant interaction with the 29 ____________________ Using the system theory, nurses view humans as open systems with many interrelated 30 _______________ When viewed in this context, a human being can be considered 31 ____________ with biologic, psychologic, social, and spiritual components. Each one of these components or subsystems can further be subdivided into smaller subsystems. Another way of looking at a human being is through holistic theory. In this theory, nurses are concerned with the individual as a 32 ____________________________ From a holistic viewpoint, all living organisms are seen as interacting, unified 33 ________________ that are more than the sum of their part.

34. Draw an example of how a System's Theory representation of a human being would look like in the box below.

Place Maslow's five categories or levels of needs in hierarchal order in the diagram below.

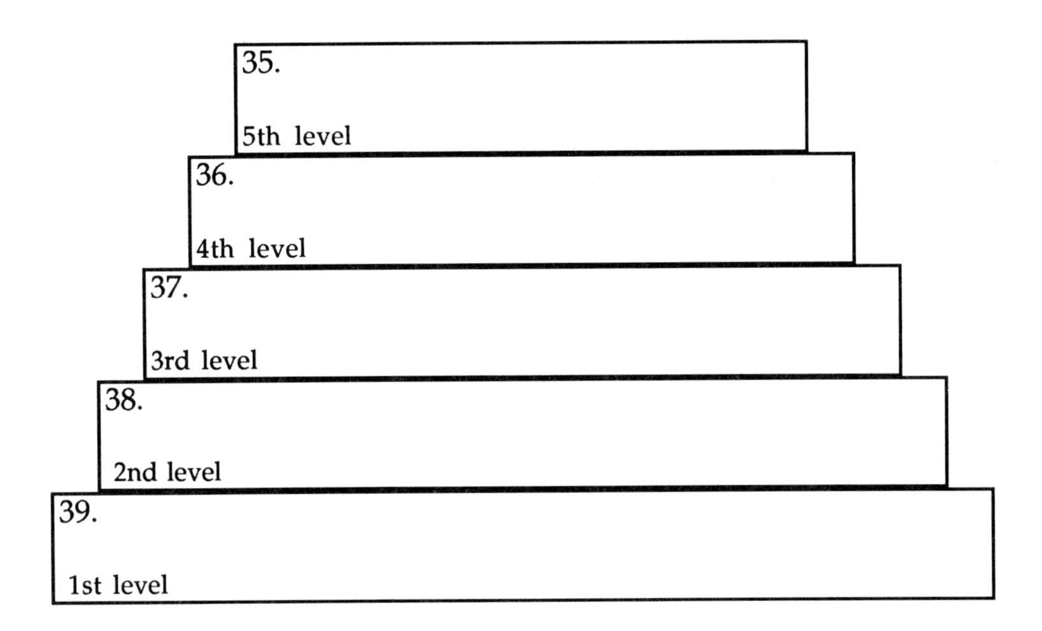

THEORIES ABOUT CARING

List and briefly describe Watson's 10 factors of caring in the space below:

40. ____

____

41. ____

____

42. ____

____

43. ____

____

44. ____

____

45. ____

____

46. ____

____

47. ____

____

48. ____

____

49. ____

____

OTHER THEORIES THAT AFFECT NURSING PRACTICE

List the 5 approaches to problem-solving:

50. ____________________

51. ____________________

52. ____________________

53. ____________________

54. ____________________

List and describe 3 phases of decision-making:

55. ____

____

____

56. ____

____

____

57. ____

____

____

Why does the nurse need an understanding of perceptual theory in order to apply the nursing process?

58. ____

____

____

____

____

____

____

____

____

Nurse Bingo: Fill in each box in the game with the appropriate person, term, theory or framework described below.

	N	U	R	S	E
1	59.	64.	69.	74.	79.
2	60.	65.	70.	75.	80.
3	61.	66.	71.	76.	81.
4	62.	67.	72.	77.	82.
5	63.	68.	73.	78.	83.

N 1 Proposed the behavioral systems model
N 2 Explains the breaking of whole things into parts
N 3 Proposed the systems interaction model
N 4 A basic structure supporting anything
N 5 Integrates concepts and statements into a configuration
U 1 Identifying a client's health problems
U 2 A set of interacting identifiable parts or components
U 3 Collecting data about a client's health
U 4 Meeting one's full potential
U 5 Concern for human attributes
R 1 A theory is devised and the research is conducted
R 2 Proposed the complementary-supplementry model
R 3 Something that is desirable, useful or necessary
R 4 Developed concept of hierarchy of needs
R 5 Proposed the science of unitary human beings model
S 1 A concept where nurses are concerned with the client as a whole
S 2 Developed ten caring factors in nursing
S 3 Developing goals and outcome criteria
S 4 Research is conducted and the findings are used to develop a heory
S 5 Proposed the self-care model
E 1 Purpose is to generate knowledge
E 2 Beliefs underlying a profession
E 3 Proposed the health care systems model
E 4 Proposed the conservation model
E 4 An abstract idea or mental image of phenomena

SELF ASSESSMENT QUESTIONS

84. A metaparadigm of nursing is

    A. a pattern for a nursing care plan that provides structure.
    B. a basic structure supporting nursing actions.
    C. a set of concepts and statements organized in a specific configuration.
    D. a set of concepts that influences the discipline of nursing.

85. "All conceptual models are frames of reference, but not all frames of reference are models." This statement is

    A. true.
    B. false.

86. If leg ulcers occur in Joseph, George and John who are mail carriers between 45 and 65 years of age, than all male mail carriers between 45 and 65 will develop leg ulcers. The preceeding statement is an example of the ______ approach in testing or developing theory.

    A. inductive
    B. deductive

87. Maslow's theory of human needs

    A. proposes that safety needs must be met before physiologic needs can be met.
    B. is based on the physiologic needs of human beings.
    C. states that needs must be completely met before the person can go on to the next level.
    D. inludes the need to know and the need to understand.

88. Which of the following types of problem solving is most useful in the clinical setting

    A. trial-and-error.
    B. intuition.
    C. experimentation.
    D. mdified Scientific.

ADDITIONAL LEARNING ACTIVITIES

1. Review the nursing care plan format on a unit in your clinical agency. What conceptual model was used to develop this plan of care? How does the conceptual framework influence how data is collected about the client in this agency?

2. Survey a group of clinical nurses in your area. Are they familiar with the conceptual frameworks used in their agency? Find out how the conceptual framework used in their agency influences their practice.

3. Review the philosophy and conceptual framework used in your educational program of nursing? How has the framework influenced the development of the curriculum and teaching strategies used ?

4. Review the various conceptual frameworks described in this chapter. Determine which framework is most similar to your own philosophy and definition of nursing.

5. After reviewing this chapter, develop your own interpretation of a nursing conceptual framework.

5 HEALTH AND ILLNESS

Interest in the quality of health and the significance of illness have gradually increased in recent years resulting in an explosion of new technology and research aimed at improving the quality and longevity of life of North Americans. Nursing has kept abreast of these changes, reemphasizing preventive care as a major focus of nursing practice. Chapter 5 focuses on the effects of illness on the client and the family during hospitalization, the issues of compliance throughout an illness and the role of the nurse in assisting the client and family through this difficult period. After completing this chapter the student will be able to:

- Differentiate health, wellness, well-being, sickness, illness, and disease.
- Identify factors that influence a person's concept of health.
- Describe how individual perceptions affect a person's health behavior.
- Describe factors affecting compliance.
- Identify nursing interventions to improve compliance.
- Describe Suchman's five stages of illness.
- Identify common behavior changes in sick persons.
- Identify effects of hospitalizations on clients.
- Describe the effects of illness on a family member's roles and functions.
- Relate current patterns and trends in health to people's nursing needs.

CONCEPTS OF HEALTH, WELLNESS, AND ILLNESS

Draw a line from the concept listed in column I to the description in column II.

1. Illness
2. Health
3. Wellness

A. One definition is: The active process of becoming aware of and making choices toward a higher level of well-being

B. A highly personal state in which the person feels unhealthy or ill

C. A state of complete physical, mental, and social well-being, and not merely the absence of disease or infirmity

List at least five of the common causes of disease.

4. ______________________________

5. ______________________________

6. ______________________________

7. ______________________________

8. ______________________________

The diagram below is a representation of factors that influence a person's concept of health. Write descriptions of these five factors in the boxes below.

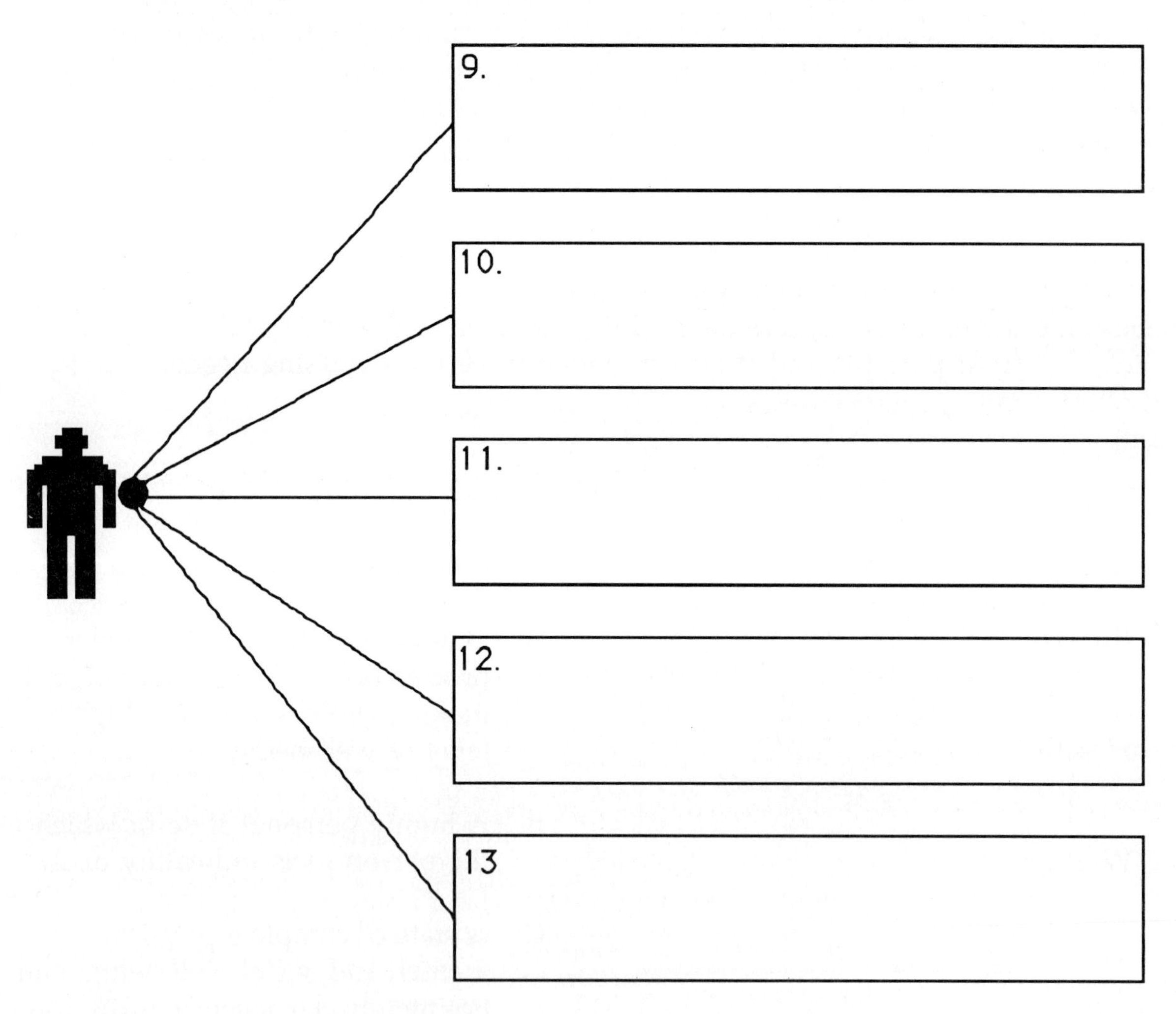

HEALTH STATUS, BELIEFS, AND BEHAVIORS

Describe how each of the following perceptions affect a person's health behavior.

14. Perceived susceptibility ______________________________

____

____

15. Perceived seriousness ______________________________

____

____

16. Perceived threat ______________________________

____

____

Complete the following statement.

Nurses preparing a care plan with a client need to consider the person's health beliefs before they attempt to change health behaviors because

17. ______________________________

____

____

Write a definition of compliance in the space below.

18. ______________________________

____

____

____

A client's degree of compliance may depend on variables such as age, education, costs, etc. In addition, complianced may be related to three factors cited in Becker's sick role model in Chapter 5. List these factors below.

19. ______________________________

20. ______________________________

21. ______________________________

Bill Riley, a 16-year-old client, was diagnosed with juvenile onset diabetes six months ago. He has been admitted to the hospital four times in the last three months because he has not followed the prescribed diet. Bill says, "I can't be bothered with this stuff. It's not fair that I have to live with diabetes for the rest of my life." What steps can Bill's nurse use to help Bill improve his compliance in the future?

22. ______________________________

23. ______________________________

24. ______________________________

25. ______________________________

26. ______________________________

ILLNESS BEHAVIORS

The following section describes Suchman's five stages of illness. Complete the following paragraph by filling in the blanks .

The first stage of illness is the 27 ______________. This is also considered the 28 ______________, when the client first believes something is wrong. This stage has three aspects: the 29 ______________ symptoms, the 30 ______________ aspect, and the 31 ______________ response. The second stage is 32 ______________. At the end of this stage, sick clients experience one of two outcomes. They may feel that the symptoms have

33 ___________________ and they feel 34 ______________________, or the symptoms increase and the client knows that 35 ___________________ should be sought. The third stage is called the 36 __________________stage. When people go for professional advice, they are really asking for the following three types of information: 37 _________________________ of real illness, 38 ______________________ of the symptoms in understandable terms, and 39 _________________ that they will be all right or prediction of the outcome. The fourth stage is the 40 ____________ client role. During this stage the client is required to give up 41 __________________. The client must go through a period of 42 ____________They require a more predictable 43 ___________. The fifth stage is the 44 _________________ stage. During this stage the client learns to give up the 45 _____________ and return to former 46 _________.

Define the term, "sick role behavior."

47. ____

____

List and describe at least four behavior changes that may be exhibited by sick persons.

48. ____

____

49. ____

____

50. ____

____

51. ____

____

Case Study:

Tony Perez is an 87-year-old Mexican-American who speaks very little English. Until recently he and his wife Ramona lived in their home of 50 years. Three months age, Ramona died suddenly. Since that time Tony has been very depressed and has shown very little interest in life. His daughter Mary took her father to the doctor when she noticed that he seemed listless and had lost weight. During the examination the doctor discovered an enlarged prostate gland and admitted Tony to the hospital for treatment. Tony had a difficult time adjusting to the hospitalization. How can Tony's nurses assist him in the following areas?

52. Loss of privacy ____

____

53. Altered autonomy ____

____

54. Altered life-style ____

____

55. Economic burden ____

____

Tony was diagnosed with cancer of the prostate gland which required surgery and chemotherapy. Even though the treatments left Tony weak and unable to care for himself, he insisted on returning home. His two daughters decided to care for Tony in his own home as long as possible. How will caring for Tony at home change his daughers's lives? Write a paragraph describing the effects of Tony's illness on his daughters.

56. ____

____

____

____

CURRENT HEALTH TRENDS

Find the pathway to health by identifying the true statements in the section that follows. Use the numbers of the true statement to guide you through the maze.

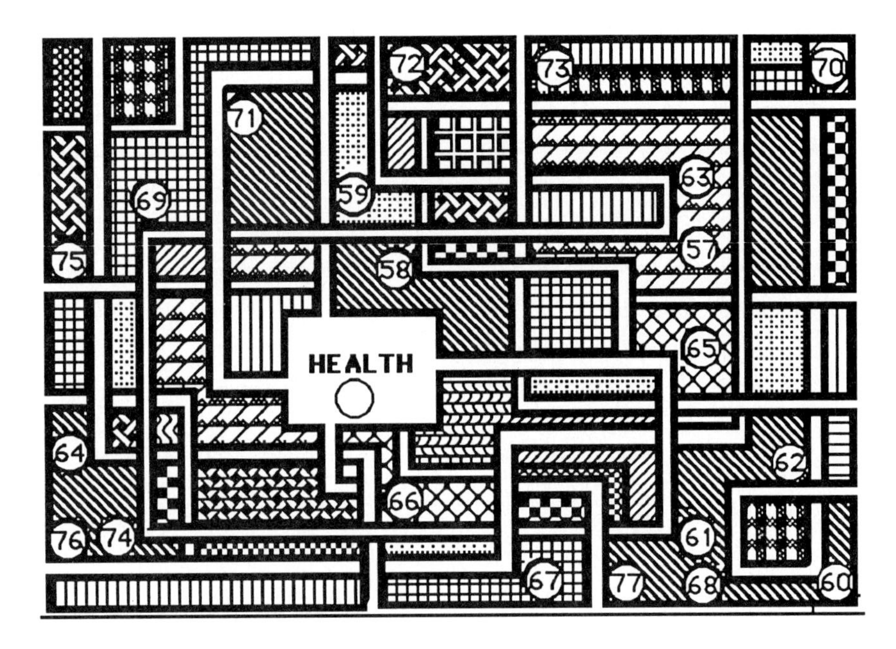

57. The number of cigarette smokers in America has declined since 1964.
58. Low serum cholesterol levels are directly related to heart attacks and strokes.
59. The infant mortality rate has declined by over 50% since 1960.
60. No significant differences exist between the life expectancies of white and black Americans.
61. Accidents are the leading cause of death for people over 35.
62. The mortality rate is the ratio of sick to well people in the population.
63. Three measures of the current health status in the United States.
64. The percentage of people who smoke 25 cigarettes or more per day has been steadily decreasing.
65. Over the past 20 years, the serum cholesterol levels of adults have declined.
66. In 1987 the life expectancy at birth for males had reached 76.5 years.
67. In 1987, life expectancy for black males was 75 years.
68. The major causes of death in infants during the first year are low birth weights and accidents.
69. The infant mortality rate has declined by over 50% since 1960.
70. Deaths from cancer have decreased over the last decade.
71. Deaths from stroke and heart disease have increased over the last decade.
72. Improvements in sanitation, housing, nutrition, and immunization have improved the health status of North Americans over the last 20 years.
73. Suicide is a major cause of death for children between 12 and 15 years of age.
74. Morbidity rate statistics for children are obtained from school records.
75. Influenza outbreaks account for one half of all absences from school.
76. The increase in cancer deaths is directly related to increases in cancer of the gastrointestinal tract.
77. In 1987, life expectancy for Hispanic women was 55 years.

THREE LEVELS OF PREVENTIVE CARE

Include a description of the three levels of preventive care in the diagram below.

78. Primary

79. Secondary

80. Tertiary

SELF ASSESSMENT QUESTIONS

81. Disease is a medical term that can be described as

   A. an alteration in body functions resulting in a reduction of capacities or a shortening of the normal life span.
   B. the perception of inbalance, lack of harmony, and loss of vitality.
   C. the inability to perform activities of daily living.
   D. the state of feeling unhealthy or ill.

82. The concept of well-being is defined as

   A. an active process of becoming aware of and making choices about health.
   B. The ability to experience physical activity, cardiovascular flexibility, and strength.
   C. a state of integrated functioning.
   D. a subjective perception of balance, harmony, and vitality.

83. The health status of an individual is

    A. the ratio between life expectancy and days of illness.
    B. the incidence of morbidity throughout the lifespan.
    C. the health of a person at a given time.
    D. the result of actions people take to improve health.

84. Which of the following statements about compliance is not true? Health compliance

    A. is the inability of a client to follow the advice of a health practitioner.
    B. can be complete, partial, or nonexistent.
    C. is related to the client's motivation to get well.
    D. is closely related to positive reinforcement by the health professional.

85. The World Health Organization proposd a definition of health in 1947 which is widely accepted. This definition states that health is

    A. a process that adapts an individual to our physical and social environment.
    B. a state of complete physical, mental, and social well-being, and not merely the absence of disease or infirmity.
    C. a dynamic state in the life cycle: illness is an interference with the life cycle.
    D. a state that is characterized by soundness or wholeness of developed human structures and of bodily and mental functioning.

ADDITIONAL LEARNING ACTIVITIES

1. Discuss the concepts of health, illness, and disease with members of your immediate family. Discuss your findings with your classmates. How do your family's views differ from the views of other families?

2. Interview an elementary schoolage child, adolescent, adult, and elderly person. How do their understanding of health, illness, and disease differ? How are they similar?

3. Interview a family of a person who is chronically ill and a family of an acutely ill client. How have these illnesses affected the family? Identify the stresses they have encountered.

6 HEALTH CARE DELIVERY SYSTEM

The health care delivery system in North America is going through enormous changes as consumer expectations become more sophistocated, and social and economic factors impact the health needs of society. The nursing profession is reacting to these changes by designing strategies that will address the health needs of the population in the future. This chapter offers the student a beginning understanding of the health related social and economic forces that are shaping the health care delivery system of the future. After completing this chapter the student will be able to:

- Describe ways in which consumers are influencing changes in health care delivery.
- Relate demographic influences to health care delivery.
- Describe the effects of specific social changes on health care delivery.
- Explain the effects of poverty and unemployment on health care delivery.
- Describe effects of the prospective payment system on health care delivery services.
- Explain how political decisions affect health care delivery.
- Compare various systems of payment for health care services.
- Describe the functions and purposes of various health care delivery systems.
- Identify the roles of various health care personnel.
- Outline the essentials of the *Patient's Bill of Rights* established by the American Hospital Association.
- Explain contemporary problems in the health care system.
- Discuss specific health care priorities and the requirements for promoting health and improving health care delivery services in the future.
- Describe the implication for nursing of changes in health care delivery systems.

HEALTH CARE SYSTEM

Define the term, health care system.

1. ______________________________

____

____

____

True and False.

2. _______ The frail aged population is the second fastest growing population in North America.

3. _______ North America is composed of people from various cultural and ethnic backgrounds which is reflected in the health care delivery system.

4. _______ About ten cents of each health care dollar in the United States are allocated to the elderly population.

5. _______ The National Health care Program in Canada ensures medical care and hospitalization for the poor and unemployed.

6. _______ Nurses should be active in issues that affect health care.

7. _______ The prospective payment system limits the amount of medicare payments made to hospitals.

8 _______ In the early 1990s, the elderly exceed twenty-eight million people and constitutes twelve percent of the population.

9. _______ By the year 2000, there will be approximately thirty-five million elderly constituting thirteen percent of the population.

10. _______ Canada is experiencing a slower rate in elderly growth population than the United States.

11. _______ Prospective payment reimbursement is made according to a classification system alled diagnostic related groups or DRGs.

FACTORS INFLUENCING HEALTH CARE SERVICES

Describe three expectations consumers have that influence the health care delivery system.

12. ____

____

____

13. ____

____

____

14. ____

____

____

FINANCING HEALTH CARE SERVICES

Describe the differance between social insurance and voluntary insurance.

15. ____

____

____

Describe the following types of insurance.

16. Medicare ______________________________________

____

17. Medicaid ______________________________________

____

18. Supplemental security income (SSI)______________________

____

19. Workman's compensation ____________________________

____

HEALTH CARE AGENCIES

Identify whether the agencies listed in the table below are primary, secondary, or tertiary providers of health care, then describe their function or service.

AGENCY	PRIMARY SECONDARY TERTIARY	FUNCTION/SERVICE
20. Home health agency		
21. Industrial clinic		
22. Crisis center		
23. Nursing home		
24. Health maintenance organization		
25. Hospice		
26. Ambulatory care center		
27. Hospital		

Complete the table below by describing the role of each health care provider listed.

HEALTH CARE PROVIDER	ROLE
28. Nurse	
29. Physician	
30. Pharmacist	
31. Social worker	
32. Dietitian	
33. Physical therapist	

RIGHTS TO HEALTH CARE

List the twelve rights of hospitalized patients in the American Hospital Association's *A Patient's Bill of Rights.*

34. ______________________________

35. ______________________________

36. ______________________________

37. ______________________________

38. ______________________________

39. ______________________________

40. ______________________________

41. ______________________________

42. ______________________________

43. ______________________________

44. ______________________________

45. ______________________________

PROBLEMS IN THE HEALTH CARE SYSTEM

Briefly describe each of the problems listed below which affect the health care system.

46. Fragmentation of care ______________________________

____

____

47. Increased cost of services ______________________________

____

____

48. Health care for the homeless ______________________________

____

____

49. Special needs of the elderly ______________________________

____

____

50. Uneven distribution of health services ______________________________

____

____

CHALLENGES FOR THE FUTURE

List five priorities for improving health care in the future.

51. ____

52. ____

53. ____

54. ____

55. ____

List five strategies nurses will need to implement in order to meet the health care needs of the future.

56. ____

57. ____

58. ____

59. ____

60. ____

SELF ASSESSMENT QUESTIONS

61. Which of the following characteristics of health care is an expectation of health care consumers today?

    A. comprehensive
    B. holistic
    C. humanistic
    D. all of the above

62. The second fastest growing population in North America is the

    A. frail aged.
    B. middle-aged "baby boomer."
    C. preschool child.
    D. well aged.

63. Which of the following is not a social insurance?

    A. Medicare
    B. Medicaid
    C. Workman's Compensation
    D. Supplemental Security Income

64. The prospective payment system

    A. reimburses hospitals according to the reasonable cost of services to the client.
    B. reimburses hospitals according to a predetermined amount for clients with a specific diagnosis.
    C. establishes categories for paying hospitals called predetermined payment system (PPS).
    D. requires that patients pay a predetermined charge for services.

65. Primary care agencies

    A. provde emergency and health maintenance care.
    B. focus on preventing complications of disease conditions.
    C. focus on rehabilitation care.
    D. include none of the above.

ADDITIONAL LEARNING ACTIVITIES

1. Review the American Hospital Association's *Patient's Bill of Rights.* How will this document affect the way you practice nursing?

2. Interview a blue collar worker, a professional person, an elderly person, middle-aged person, and a college student. Investigate their expectations of health care for the next ten years. Report your findings in class.

3. Research the problems of the homeless in your community. What are their health needs? How is your community responding to their needs?

4. Interview an elderly person who has been hospitalized recently and who feels comfortable discussing his or her Medicare reimbursement. Has the prospective payment system provided sufficient reimbursement to cover the client's hospital bill? Does the client have difficulty completing the required paperwork that is necessary for Medicare and insurance reimbursement? Report your findings back to the class.

7 ETHICAL ASPECTS OF NURSING PRACTICE

The caring nature of nursing requires the nurse to assume a set of values, beliefs, and ethics that reflects the human dignity of the client. The first step in acquiring these values is to review and clarify the belief systems the nurse already holds. This chapter provides the learner with an overview of how values, beliefs, and codes of ethics affect the practice of nursing, the care-giver, and the client. After completing this chapter the student will be able to:

- Identify various ways in which values, beliefs, and attitudes are learned.
- Describe four methods of transmitting values.
- Explain Raths' values clarification process.
- Describe reasons for identifying clients' values.
- State three elements of a moral dilemma.
- Identify professional values incorporated in nursing codes.
- Discuss the purposes of codes of ethics.
- Discuss the types of ethical issues encountered by nurses.
- Discuss the process for resolving ethical dilemmas.

VALUES, BELIEFS, ATTITUDES, AND ETHICS

In the spaces provided, match each term in column I with the appropriate term in Column II.

1. ______ Values
2. ______ Ethics
3. ______ Intrinsic value
4. ______ Attitudes
5. ______ Belief
6. ______ Value system

A. Originates outside an individual

B. Organization of a person's values along a continuum

C. Relates to the maintenance of life

D. Principles that govern conduct

E. Something accepted as true by a judgment or probability

Define the following terms:

7. Bioethics ______________________________

____

____

8. Morality ______________________________

____

____

9. Moral reasoning ______________________________

____

____

10. Value set ______________________________

____

____

VALUES

List at least three ways in which values, beliefs, and attitudes are transmitted:

11. ______________________________

12. ______________________________

13. ______________________________

Read the descriptions of value transmission described in each box below, then identify the method used in the space provided:

A group of scouts learns to care for the environment by following the example of their scout leader. This is an example of: 14. ______________________	After hitting a friend, a child is told, "You can stay in your room for the rest of the day, or you can apologize." This is an example of: 16. ______________________
A 10-year-old boy harasses a neighbor's dog. The neighbor tells the police that the child's parents are never home. This is an example of: 15. ______________________	Mrs Conway tells her 16-year-old daughter that she will go to hell if she lets a boy "take advantage" of her. This is an example of: 17. ______________________

18. Define the term "values clarification:"

____

____

____

List the 7 parts of the values clarification process:

19. ______________________ 23. ______________________

20. ______________________ 24. ______________________

21. ______________________ 25. ______________________

22. ______________________

Case Study:

Marvis Thompson is an RN who works in a communicable disease unit in a large midwest city. A meeting was called for all staff and personnel to discuss upcoming changes that were planned for the unit. The assistant administrator announced that the unit's focus was being changed to care for clients with AIDS. Marvis and her co-workers were very upset. Some people resigned immediately or asked for transfers to other departments. Some of the actions Marvis took are described below. Identify which of the seven values clarification steps each of the actions represents by drawing a line from the action to the appropriate step.

ACTIONS	VALUE CLARIFICATION STEPS
26. At the meeting, Marvis stood up and said, "I don't believe we should discriminate between clients who have AIDS and other clients with communicable diseases.	Prizing and cherishing
27. Marvis continued to work on the communicable disease unit with AIDS clients even though many of the former employees resigned.	Publicly affirming when appropriate
28. Marvis stated, "I realize there is a slight risk of contacting the disease if I am not careful, but I fully understand what I am doing.	Choosing from alternatives
29. One of Marvis' co-workers said, "Marvis has always felt that it was a privilege to work with clients who were considered poor, undesirable, or socially unacceptable.	Choosing after consideration of consequences
	Choosing Freely
	Acting
	Acting with pattern, consistency and with repetition

List and describe four reasons why a nurse needs to identify a client's values:

30. ______________________________

____

31. ______________________________

____

32. ______________________________

____

33. ______________________________

____

Mrs Rose Shultz, a 55 year-old school teacher, has been diagnosed with advanced myeloma, a potentially terminal disease. Her doctors have told her that her life could be prolonged if she would agree to a course of chemotherapy. When you discuss her treatment plan with her, she indicates that she understands her life may be prolonged, but she wonders whether the side effects of the treatment are worth the extra time it will give her. How can you, Mrs. Shultz's primary nurse, help her to decide whether she should have this treatment or not? Use the seven steps of value clarification to formulate your answer.

34. ______________________________

____

____

ETHICS

Although the terms *ethics* and *morals* are often used interchangeably, they do have very different meanings. Differentiate between the two terms:

35. ______________________________

____

In the case study cited earlier in this chapter, Marvis Thompson was faced with caring for clients with acquired immune deficiency syndrome (AIDS). One of Marvis'co-workers, Joyce Jackson, RN, told Marvis that she would refuse to care for any client with AIDS. Marvis explained to Joyce that the ANA had developed four fundamental criteria to determine whether nurses had the ethical and moral obligation to care for a client. List the four criteria in the space below.

36. ____

____

37. ____

____

38. ____

____

39. ____

____

Joyce gradually worked through her fears concerning AIDS and agreed to work on the unit. Soon after, she was assigned to be the primary nurse for John Riley, a 24 year-old man with AIDS. John was in the advanced stages of the disease and had been told by his doctor that his prognosis was poor. When Joyce talked to John later that day he said, "I want the doctor to do everything possible to save my life." When Joyce related John's message to the doctor, he said, "John is going to die no matter what we do. We're just going to keep him comfortable until the end." Why is this situation a moral dilemma for Joyce? State the three criteria that determine whether a situation is a moral dilemma.

40. ____

____

41. ____

____

42. ____

____

Describe the six steps that Joyce can take to solve this moral dilemma.

43. ____

____

44. ____

____

45. ____

____

46. ____

____

47. ____

____

48. ____

____

Following are several alternative actions that Joyce decided she could take in working through this moral dilemma. Read each course of action that Joyce could take, then describe the possible outcomes(s) or consequence(s) of that action in the space provided.

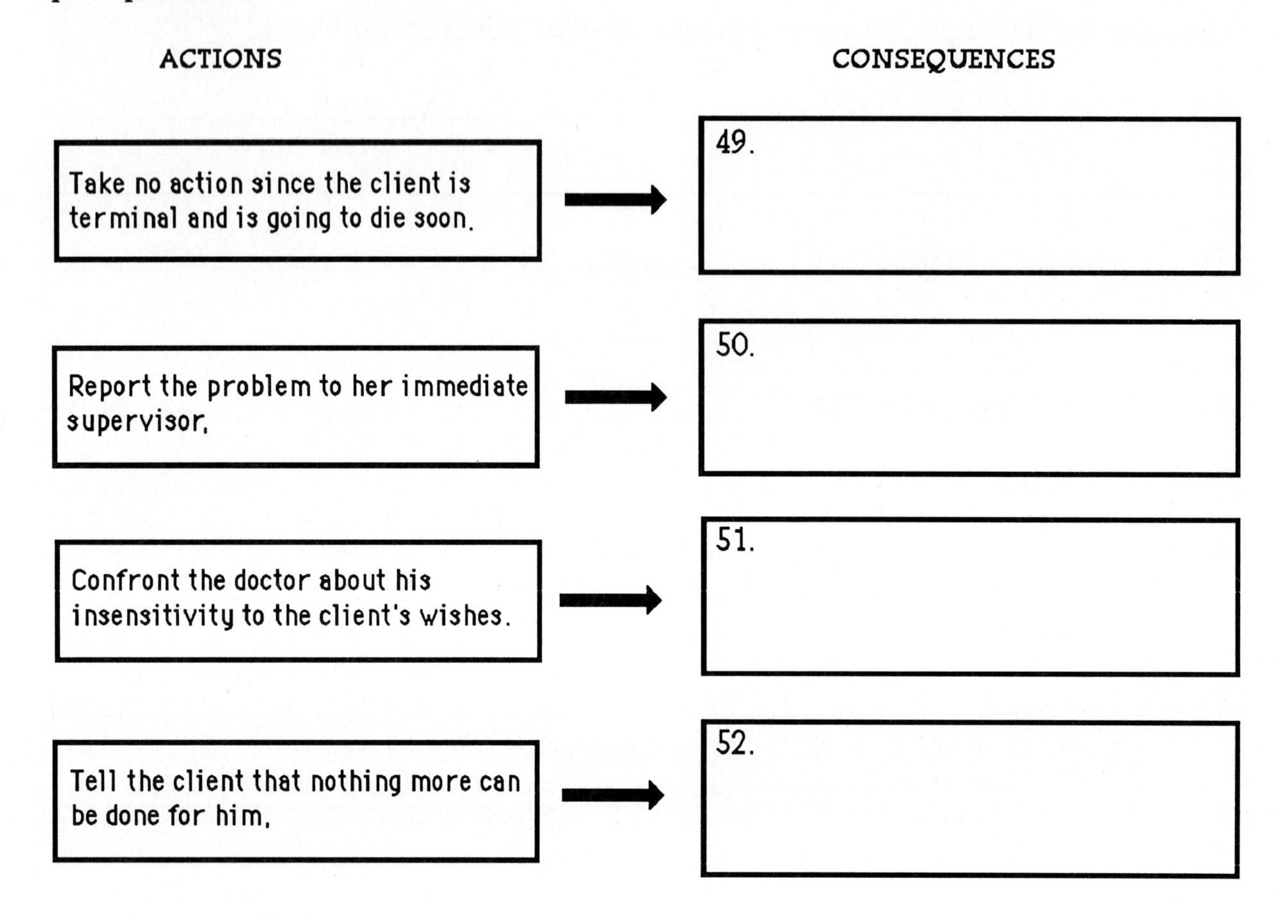

SELF ASSESSMENT QUESTIONS

53. Codes of ethics

    A. are shared by the persons within the profession.
    B. give the general public a frame of reference for judgements in complex nursing situations.
    C. provide specific rules that all nurses must follow.
    D. are legislated in each state as part of the nurse practice act.

54. The best way to assess a client's values is to

    A. talk to the client, family, and significant others.
    B. review health records and past history.
    C. give the client a values clarification test.
    D. all of the above.

55. An extrinsic value

    A. can be either positive or negative.
    B. originates outside the individual.
    C. is necessary for the maintenance of life.
    D. explains how something should be.

56. Morality concerns behavior which involves judgements, actions, and attitudes based on rationally conceived and effectively established norms. This statement is

    A. true.
    B. untrue.

57. After being oriented to the care of clients with AIDS, Joyce Jackson, RN, said, "After researching AIDS and completing this workshop, I'm feeling much more positive about caring for these clients." Which of the following criteria supports the conclusion that Joyce's values concerning client's with AIDS have changed? Her remarks demonstrate that she made this decision

    A. without pressure from others.
    B. after reflecting on her choices.
    C. with a willingness to share her thinking with others.
    D. based on all of the above factors.

ADDITIONAL LEARNING ACTIVITIES

1. Answer the following questions: What ten things do I like to do? What ten things do I like about myself? What ten things do I dislike about myself? Share

and discuss your results with a friend and/or classmate. Will the results of the insights that you have learned cause you to change any aspect of your life?

2. Complete the values clarification exercises that follow this section. Keep in mind that there are *no* wrong answers. Discuss your results with a friend or in class.

3. After completing the above activities, develop a list of ten values that you will use to guide your daily interactions or activities.

4. Compare the codes of ethics of the Canadian Nurses' Association, the International Council of Nurses, and the American Nurses' Associations. What are the similarities? What are the differences?

5. Ask a nurse in the clinical setting to describe how the code of ethics for nurses affects his or her practice.

VALUE CLARIFICATION EXERCISE 1

SA= Strongly agree, U = Undecided,SD = Strongly disagree, A = Agree, D= Disagree

	BELIEFS	SA	A	U	D	SD
SEX	Teenage sex is acceptable.					
	Casual sex is acceptable.					
	Sex is an expression of love and commitment.					
	Group sex is acceptable.					
	Clients should make decisions about own health care.					
RIGHT TO DIE	Clients have a right to refuse extraordinary life-sustaining treatment.					
	Refusing life-sustaining treatment is a form of suicide.					
	Clients have a right not to be interfered with in a rational act of suicide.					
	Comfort measures should always be provided.					
ABORTION	A woman should have the right to choose abortion.					
	Abortion at any point during gestation is murder.					
	Abortion should be performed if the woman's health is endangered.					
	A mentally handicapped woman should be encouraged to have an abortion.					
HEALTH	A nurse should be a role model of health.					
	An obese nurse can effectively instruct an obese client about nutrition and exercise.					
	A nurse who smokes can help a client stop smoking.					
HEALTH CARE	A nurse's major accountability is to the physician.					
	A nurse's major accountability is to the institution.					
	A nurse's major accountability is to the client.					
	Both client and nurse have decision-making responsibility.					

VALUE CLARIFICATION EXERCISE 2

Read the story below, then anwer the questions which follow.

Tricounty Medical Center has had a crisis. A helicopter crashed, hitting the northeast wing of the building where the delivery room and nursery are located. In the midst of the explosion, fire and confusion, five people make their way to the roof of the 4th floor wing and wait for rescue. They are Mary, who is carrying her newly-delivered baby; Peter, her husband; Jake, MD, an obstetrician who delivered Peter and Mary's baby; Gloria, RN, who was on duty in the delivery room; and Morgan, a pharmaceutical salesman who was in the OB department delivering drug samples.

Mary says to the group, "I think I hear someone crying! Yes, I can hear cries for help. We've got to go back down the stairs and help those people!" Dr. Jake says, "You women stay here. Peter and I will go back down and see what we can do." Peter looks first at the baby then down the smoke-filled stairs and replies, "I'm not going back down. It's too risky. The smoke is too thick. We'd never make it through and survive." Gloria the RN says, "We've got to do something! We can't just let people die. I'd never be able to live with myself. Those are my patients and I need to help them!" Gloria runs down the stairs and disappears. Morgan says, "You're all crazy! In a situation like this it's every man for himself!"

Now that you've read the story, answer the following questions. Remember there are *no* wrong answers. Give a rationale for each of your answers.

1. Which of the characters in the story do you most admire? Least admire?

2. Was Peter selfish in not wanting to risk his life?

3. Was Morgan's behavior unethical?

4. Was Gloria's decision to return to her clients foolish?

VALUE CLARIFICATION EXERCISE 3

Apply the steps for resolving an ethical dilemma to each of the following situations:

Situation 1: Mattie Masters is a 25-year-old client with cerebral palsy who lives in a halfway house where she shares an apartment with two other girls who are also handicapped. You are the nurse who visits Mattie once a month to assess her health needs and provide necessary care. On one such visit, Mattie tells you that she wants to take birth control pills since she has become sexually active with a male resident next door. You know that Mattie has a blood disorder that contraindicates the use of birth control pills.

Situation 2: You are working in the gynecological unit of a large metropolitan hospital where you are assigned to Joan Wylie, a 16-year-old girl with a diagnosis of leukemia. She is scheduled for a tubal ligation. On the day before surgery she tells you, "I don't want to have this surgery but I have no choice. My mother signed the papers." Joan's mother tells you, "I had to consent to save her life. Joan is sexually active and the doctors have told me Joan could die if she gets pregnant. What else could I do?"

Situation 3: Mr Morgan is a 45-year-old man with severe congestive heart failure. His doctors have told him that his life could be prolonged by a new drug that is being tested at the university hospital where he is a client. Although the side effects can be very severe, he could live several years longer. When you discuss this new drug treatment with the client he says, "Why should I take a drug that will make my life miserable? I'd rather die now!" His wife becomes very upset when Mr Morgan tells her his decision. Later she tells you, "If he doesn't agree to the treatment, I'll have him declared incompetent and I'll sign for him!"

8 LEGAL ASPECTS OF NURSING PRACTICE

More than ever before in the history of nursing, the nurse must be aware of the legal contraints underlying professional practice. The heightened awareness of the consumer, increased technology in medical procedures, and the expanded role in nursing are some of the factors that make nurses more vulnerable to legal action. This chapter presents an overview of the legal concepts the nurse needs to protect the client, the employer and him- or herself. After completing this chapter the student will be able to:

- Describe general legal concepts as they apply to nursing.
- Explain how nurse practice acts legally help the nurse practitioner.
- Differentiate mandatory from permissive licensure.
- Describe ways standards of care, agency policies, and nurse practice acts affect the scope of nursing practice.
- Identify ways nursing students can minimize chances of liability.
- Identify essential types and elements of contracts.
- Identify rights and obligations associated with the nurse's legal roles.
- Describe collective bargaining with reference to nursing.
- Identify areas of potential liability for nurses.
- Differentiate crimes from torts and give examples in nursing.
- Describe the purpose and essential elements of informed consent.
- List information that needs to be included in an incident report.
- Describe actions a nurse should take when a client is injured.
- Explain the positive and negative aspects of living wills.
- Describe the purpose of professional liability insurance.

GENERAL LEGAL CONCEPTS

The term law can be defined as:

1. ______________________________

____

____

____

Briefly describe the four functions the law serves in nursing.

2. ____

____

3. ____

____

4. ____

____

5. ____

____

Identify and describe the various types of the law by completing the diagram below.

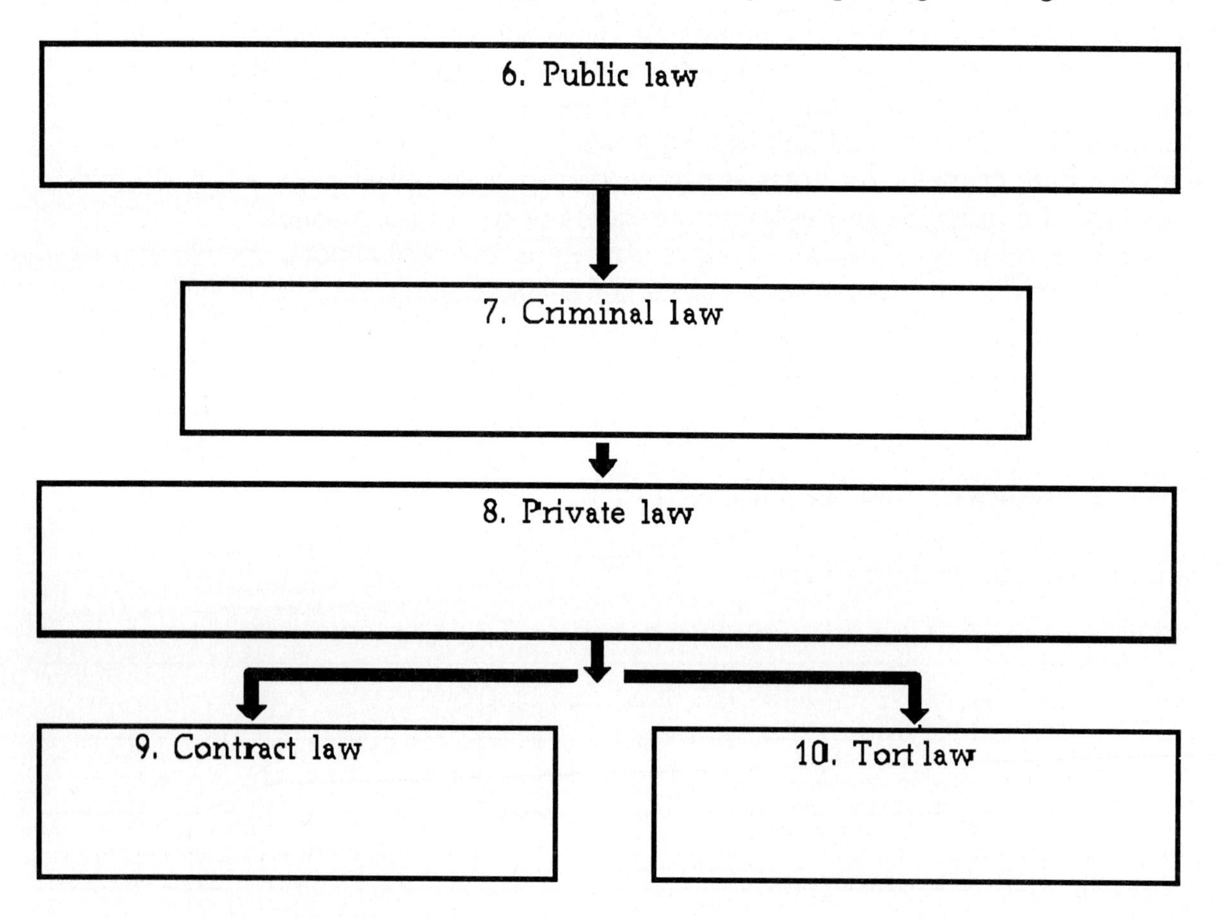

LEGAL ASPECTS OF NURSING

Identify each of the following descriptions in the space provided.

11. ____________________ is the listing of an individual's name and other information on the official roster of a governmental or nongovernmental agency.

12. ____________________ is a process by which a voluntary organization or governmental agency appraises institutions, programs, or services.

13. ____________________ protects the nurse's professional capacity and legally controls nursing practice through licensing.

14. ____________________ is the process of determining and maintaining competence in nursing practice.

15. ____________________ are legal permits granted by a government agency to individuals to engage in the practice of a profession.

16. ____________________ is the voluntary practice of validating that an individual nurse has met minimum standards of nursing competence in specialty areas.

CONTRACTUAL ARRANGEMENTS IN NURSING

Briefly describe how each of the following factors affect the scope of nursing practice.

17. Standards of care ____

____

____

18. Nurse practice acts ____

____

____

19. Agency policies ______________________________

____

____

List the four elements that must be met to make a contract valid.

20. ______________________________

21. ______________________________

22. ______________________________

23. ______________________________

Complete the following statement:

An organized strike could present a moral dilemma for a professional nurse because:

24. ______________________________

____

____

____

POTENTIAL LIABILITY AND SELECTED LEGAL FACETS IN NURSING

The scramblegram below contains legal terms related to nursing practice discussed in Chapter 8. Identify each term in the definition section below, then circle the term in the scramblegram. The words may appear horizontally, vertically, diagonally, backward, or forwards.

a	w	n	e	r	s	l	o	w	c	o	d	e	a	t	h	t	y
q	s	o	r	g	a	n	d	o	n	a	t	i	o	n	u	i	o
p	d	i	a	s	d	f	q	e	c	n	e	q	i	l	g	e	n
n	f	t	h	j	j	j	k	l	l	l	z	q	x	c	v	a	b
m	g	r	d	s	d	u	a	r	f	e	l	o	n	y	e	g	h
q	w	o	e	e	e	e	c	v	a	f	r	r	q	m	j	m	b
h	u	b	f	u	o	p	n	b	l	v	b	w	e	c	g	t	y
z	d	a	a	t	e	r	w	b	s	a	u	d	i	k	m	b	v
a	m	s	m	h	c	r	i	m	e	f	s	l	a	n	d	e	r
a	a	d	a	a	o	e	l	r	i	i	v	s	n	d	q	e	d
u	l	f	t	n	n	x	l	b	m	r	k	p	a	q	w	e	r
t	p	g	i	a	s	k	l	n	p	c	i	n	q	u	e	s	t
o	r	h	o	s	e	e	c	o	r	o	n	e	r	m	l	n	b
p	a	j	n	i	n	a	s	d	i	f	g	h	j	k	z	t	z
s	c	k	h	a	t	m	a	n	s	l	a	u	g	h	t	e	r
y	t	l	h	j	k	l	a	x	o	t	i	g	c	y	h	n	m
c	i	n	f	o	r	m	e	d	n	w	e	b	a	t	t	e	r
v	c	m	c	v	i	t	i	i	m	s	s	d	e	x	d	f	g
b	e	n	i	t	r	f	r	r	e	d	r	e	c	l	c	g	v

25. ______________ A declaration by a person about how his or her property is to be disposed of after death

26. ______________ An agreement by a client to accept a course of treatment

27. ______________ The act of painlessly putting to death persons suffering from incurable disease

28. ______________ An act committed in violation of public law

29. ______________ To omit doing something that a reasonable person would do

30. ______________ No effort is to be made to resuscitate the client

31. _____________ A crime of a serious nature such as murder

32. _____________ Unjustifiable detention

33. _____________ A threat to touch another unjustifiably

34. _____________ A "half-hearted" resuscitation effort is to be made

35. _____________ An examination of the body after death

36. _____________ Second degree murder

37. _____________ Untrue words that damage one's reputation

38. _____________ A legal inquiry into the cause of a death

39. _____________ Defamation by print, writing, or pictures

40. _____________ A public official who inquires into the causes of death when appropriate

41. _____________ An offense of a less serious nature

42. _____________ A gift of all or any part of the body for education, research, or transplantation

43. _____________ Communication that is false

44. _____________ Recognition of the client's right to refuse life-saving treatment

45. _____________ A civil wrong committed against a person

List the six types of information needed for an incident report.

46. ____

47. ____

48. ____

49. ____

50. ____

51. ____

LEGAL PROTECTIONS FOR NURSES

Why is it important that the practicing professional nurse carry malpractice insurance?

52. ____

SELF ASSESSMENT QUESTIONS

53. Tort law includes all of the following except

    A. negligence.
    B. malpractice.
    C. Invasion of privacy.
    D. homicide.

54. Student nurses who practice nursing as part of their education must have

    A. mandatory licensure.
    B. permissive licensure.
    C. both of the above.
    D. neither of the above.

55. The professional nurse

    A. is expected to provide safe and competent care.
    B. has legal rights that are the same as any other citizen.
    C. has the right to engage in collective bargaining.
    D. all of the above.

56. Which of the following is not an essential element of informed consent?

    A. The consent must be given voluntarily.
    B. The consent must be given by an individual with the capacity and competence to understand.
    C. The client must be given assurances of complete recovery.
    D. The client must be given enough information to be the ultimate decision maker.

57. If a client is injured, nurses must take steps to protect

    A. the client.
    B. themselves.
    C. the employer.
    D. All of the above.
    E. A and E.

ADDITIONAL LEARNING ACTIVITIES

1. Review a copy of the nurse practice act in your state. Identify the areas of the act that relate to independent and dependent practice.

2. Obtain a copy of a policy or brochure that describes malpractice insurance. Determine the value to a nurse practitioner.

3. Collect magazine and newspaper articles concerning right-to-life issues in the media. Bring them to class for discussion.

9 INTRODUCTION TO THE NURSING PROCESS

In ancient times, nurses provided nursing care that largely depended on unvalidated intuitive knowledge handed down from one generation to another. Today a body of nursing knowledge derived from theory development, research, and practice has developed, which gives validation and forms the foundation for modern nursing practice. In the past thirty years, the nursing process, a systematic, rational method for implementing that body of nursing knowledge, has evolved. Since the nursing profession adopted this unique method, the quality of nursing care has increased, clients have became participants in their own care, and nursing has become more accountable. After completing this chapter the student will be able to:

- Describe the components of the nursing process.
- Identify the contribution of selected nurses to the development of the nursing process.
- Identify nursing activities involved in each component of the nursing process.
- Identify essential characteristics of the nursing process.
- List benefits of the nursing process to the client.
- List benefits of the nursing process to the nurse.
- Describe how the nursing process is a framework for accountability and responsibility.

HISTORICAL PERSPECTIVES

True and False

1. _______ Before the nursing process was developed, nurses provided care that was based on the medical orders written by physicians.

2. _______ The concept of the nursing process has been used since the early 19th century.

3. _______ Wiedenbach described the nursing process as three steps, including observation, ministration of help, and validation.

4. _______ The five steps of the nursing process which are commonly used today were first published by the American Nurses' Association in the *Standards of Nursing Practice.*

5. _______ The five steps of the nursing process described in *Standards of Care* in 1973 were derived from the nurse practice acts that had been legislated in several states.

6. _______ The Western Interstate Commision on Higher Education identified the steps of the nursing process as perception, communication, interpretation, intervention, and evaluation.

7. _______ The term "analyzing" as defined by the National Council of State Boards of Nursing is used in the same context as the term "diagnosing."

8. _______ Nursing is the diagnosis and treatment of human responses to actual or potential health problems.

COMPONENTS OF THE NURSING PROCESS

Complete the following nursing Process table by filling in the purpose(s) and related activities for each component.

COMPONENTS AND PURPOSE(S)	ACTIVITIES
Assessing 9.	10.
Diagnosing 11.	12.
Planning 13.	14.
Implementing 15.	16.
Evaluating 17.	18.

List the ten characteristics of the nursing process in the space below:

19. ______________________________

20. ______________________________

21. ______________________________

22. ______________________________

23. ______________________________

24. ______________________________

25. ______________________________

26. ______________________________

27. ______________________________

28. ______________________________

BENEFITS OF THE NURSING PROCESS

The advantages the nursing process offers the nurse and the client are listed below. Explain each of the benefits or advantages in the space provided.

Benefits For the Client:

29. Quality client care ______________________________

____

30. Continuity of care ______________________________

____

31. Participation by clients in their health care ______________________________

____

For the Nurse:

32. Consistent and systematic nursing process ______________________________

____

33. Job satisfaction ____

____

34. Professional growth ____

____

35. Avoidance of legal action ____

____

36. Meeting professional nursing standards ____________________________

____

37. Meeting standards of accredited hospitals __________________________

____

A FRAMEWORK FOR ACCOUNTABILITY

Discuss how the nursing process provides a framework for accountability in professional nursing.

38. ____

____

____

____

SELF ASSESSMENT QUESTIONS

39. The group responsible for the development of the steps of the nursing process is the

    A. National League for Nursing.
    B. National Council of State Boards of Nursing.
    C. Western Interstate Commission on Higher Education.
    D. American Nurses' Association.

40. The nursing process can be defined as

    A. a systematic, rational method of planning and providing nursing care.
    B. the conclusion or judgement which occurs as a result of nursing assessment.
    C. a series of planned actions or operations directed toward a particular result.
    D. a systematic method for providing nursing care under the direction of a physician.

41. The first national conference on the classification of nursing diagnoses in 1973 was initiated by

    A. Heidgerken and McCain.
    B. Henderson and Wiedenbach.
    C. Peplau and Orem.
    D. Gebbie and Lavin.

42. The term *nursing intervention* may be used interchangeably with the term *implementation*.. This statement is

    A. true.
    B. false.

43. Accountability in professional nursing is

    A. being responsible for one's professional actions.
    B. implemented through the nursing process framework.
    C. includes both of the above.
    D. A only.

ADDITIONAL LEARNING ACTIVITIES

1. Apply the five phases of the nursing process to a problem in your life. How did the steps help you solve your problem?

2. Interview a nurse who uses the nursing process in client care. How does the nurse incorporate the process in daily nursing care?

10 ASSESSING

Assessing is a continuous process that is initiated at the beginning of the nurse-client relationship and carried out through each phase of the nursing process. Without the process of assessment the nurse would be unable to establish a nursing diagnosis, plan and implement nursing interventions, or determine whether outcome criteria have been accomplished. Assessment is the foundation for efficient professional nursing practice. After completing this chapter the learner will be able to:

- Identify the purposes of assessing.
- Differentiate objective and subjective data and variable and constant data.
- Identify three methods of data collection.
- Identify the purposes of each method of data collection and give examples of how each is useful.
- Compare closed and open-ended questions, providing examples and listing advantages and disadvantages of each.
- Describe important aspects of the interview setting.
- Identify four techniques used during the physical examination.
- Explain common health areas the nurse assesses.
- Contrast various frameworks used for nursing assessment.
- Describe the importance of assessing to nursing diagnosis.
- Describe the importance of reassessing during other phases of the nursing process.

PURPOSE OF ASSESSING

List the reasons why assessing is important in each of the following phases of the nursing process.

1. Diagnosis phase ______________________________

2. Planning and implementing phases ______________________________

3. Evaluation phase ______________________________

Define the following terms:

4. Data base ____

____

____

5. Data collection ____

____

____

6. Premature closure ____

____

____

TYPES OF DATA

Identify the types of data described below by placing an SD for subjective data, an OD for objective data, and a VD for variable data in the space provided. There may be more than one category for some of the data.

7. ________ Blood pressure

8. ________ Vomiting

9. ________ Temperature

10. ________ Double vision

11. ________ Swollen ankles

12. ________ A headache

13. ________ A backache

14. ________ Blood in the urine

15. ________ Bruised area

16. ________ Tremors

SOURCE OF DATA

Complete the secondary source table that follows by describing each secondary source listed, then list the information each source provides in the assessing phase.

DESCRIPTION OF SECONDARY SOURCES	INFORMATION PROVIDED
Significant others 17.	18.
Health personnel 19.	20.
Medical records 21.	22.
Literature 23.	24.

METHODS OF DATA COLLECTION

Describe the three methods of data collection.

25. Observing ______________________________

____

26. Interviewing ______________________________

____

27. Examining ____

____

Complete the methods of data collection table below, identifying the purposes of each method. Include an example of how each method is useful in assessing the client.

METHODS OF DATA COLLECTION PURPOSE	EXAMPLE
28.	29.
30.	31.
32.	33.

Identify the following questions as "closed" or "open" in the space provided.

34. _______ When did the symptoms first appear?

35. _______ When did you fall?

36. _______ What brought you to the clinic?

37. _______ Are you experiencing any pain now?

38. _______ How do you feel about staying another day in the hospital?

39. _______ Are you tired?

The interview and setting are an important part of the assessment phase. In the space below describe each aspect of the interview setting.

40. Time ____

____

____

____

41. Place ____

____

____

____

42. Seating arrangement ______________________________

____

____

____

List the four techniques the nurse uses during the physical examination.

43. ____

44. ____

45. ____

46. ____

STRUCTURING DATA COLLECTION

Most nursing health histories are organized around a model or framework developed by a theorist with a nursing and/or psychosocial background. In the exercise that follows, match the theorist in column I with the assessment method in column II.

	I	II
47. _______	Gordon	A. Seven category /intervention tool
48. _______	Roy	B. Eleven functional health patterns
49. _______	Newman	C. Hierarchy of needs
50. _______	Piaget	D. Eight universal self-care requisites
51. _______	Maslow	E. Assessment of cognitive development
52. _______	Orem	F. Four adaptive modes
53. _______	Selye	G. Stress theory

SELF ASSESSMENT QUESTIONS

54. Which of the following statements about assessing is not true? Assessing

    A. is only done during the first phase of the nursing process.
    B. must be done before a nursing diagnosis can be formulated.
    C. can be used to establish a database about a client's response to health concerns or illness.
    D. involves data collection and validation.

55. Which of the following comments about the client's fluid intake would be most appropriate to include in the data base of a client who is dehydrated? The client

    A. is drinking frequently.
    B. asks for fluids periodically.
    C. prespires often.
    D. drank 8 glasses of water.

56. Which of the folowing questions is most likely to elicit meaningful information about nausea?

    A. "Is the nausea related to eating?"
    B. "What were you doing when the nausea started?"
    C. "Do you experience pain when the nausea starts?"
    D. "Describe how you were feeling when the nausea started."

57. An advantage of using open questions during an assessment interview is

    A. they require less effort from the client.
    B. they take less time.
    C. responses are easily documented.
    D. they let the client do the talking.

58. Assessing is important in developing a nursing diagnosis because

    A. it validates the diagnosis.
    B. it determines the outcomes of the nursing strategies.
    C. it involves the participation of the client.
    D. it includes all areas of human response.

ADDITIONAL LEARNING ACTIVITIES

1. Review the assessment tool used in your clinical setting. What nursing model is used as an organizing framework for the tool?

2. Use the tool you reviewed in the previous activity to interview a client. Was the tool helpful? Confusing? Discuss your responses.

3. Practice using various distances when conversing with others. What distances were most conducive for conversation?

4. While providing nursing care to a client, practice using open and closed questions. Which technique elicited the most information from the client?

11 DIAGNOSING

Developing the art and science of diagnosing a client's response to health problems has given nursing a sharper focus for client care and a clearer direction for planning independent nursing actions. Since nursing diagnosis has been established as the recognized method for identifying client problems, the practice of nursing has become more client centered, the nursing process has been strengthened, and the quality of care has improved. This chapter will reinforce the value of nursing diagnosis as part of the nursing process. After completion of this chapter the student will be able to:

- Define the term *nursing diagnosis.*
- Compare medical and nursing diagnoses.
- Identify basic steps in the diagnostic process.
- Describe the PES format for writing nursing diagnoses.
- Describe the characteristics of a nursing diagnosis.
- List common errors in writing diagnostic statements.
- Describe the evolution of the nursing diagnoses movement.
- List advantages of a taxonomy of nursing diagnoses.
- Identify the challenges of the profession related to nursing diagnoses.

NURSING DIAGNOSIS

True and False

1. _______ The word diagnosis is derived from the Greek word diagignoskein, which means "to distinguish."

2. _______ Nurses are not educated to diagnose or treat diseases.

3. _______ A nursing diagnosis is a statement of nursing judgement.

4. _______ The term *diagnosis* may be defined as a process of analysis and synthesis

5. _______ Physicians tend to perform the diagnostic process more efficiently than nurses.

6. _______ Nursing diagnosis may describe either actual or potential health problems.
7. _______ A health problem is any condition or situation in which a client requires help.

8. _______ The first National Conferance on the Classification of Nursing Diagnosis was held in 1973.

9. _______ A nursing diagnosis usually consists of a four part statement.

10. _______ Nursing diagnoses help nurses focus on dependent nursing actions.

A list of medical and nursing diagnoses follow. Identify the medical diagnoses with an M and the nursing diagnoses with an N.

11. _______ Appendicitis

12. _______ Pneumonia

13. _______ Powerlessness

14. _______ Alteration in comfort

15. _______ Dysfunctional grieving

16. _______ Multiple sclerosis

Describe the difference between an actual and a potential health problem in the space below.

17. ____

____

____

____

THE DIAGNOSTIC PROCESS

Identify and briefly describe each step of the diagnostic process in the diagram below.

18.
19.
20.

THE DIAGNOSTIC PRODUCT

Identify and describe each part of a nursing diagnostic statement using the PES format.

21. P ____

22. E ____

23. S ____

Check the incorrectly stated nursing diagnoses in the list below.

24. ______ Appendicitis related to inflamed appendix resulting in surgical intervention

25. ______ Alteration in comfort related to pain and swelling resulting in lack of sleep

26. ______ Ineffective breathing patterns related to industrial pollution resulting in coughing, shortness of breath and chest pain

27. ______ Ineffective airway clearance resulting in cyanosis caused by mucus resulting in excessive drainage

28. ______ Unable to sleep resulting in fatigue

29. ______ Ineffective coping related to death of father resulting in inability to concentrate

30. ______ Hernia repair related to abdominal pain resulting in surgical intervention

31. ______ Self-care deficit related to immobilized ankle resulting in inability to walk

THE NURSING DIAGNOSIS MOVEMENT

List at least three advantages a taxonomy of nursing diagnoses has for nursing.

32. ____

33. ____

34. ____

SELF ASSESSMENT QUESTIONS

35. A nursing diagnosis is different from a medical diagnosis in that the medical diagnosis

    A. guides independent nursing action.
    B. consists of a two part statement.
    C. describes an individual's response to a disease.
    D. remains constant throughout the duration of the illness.

36. Nursing diagnoses are valuable to nursing practice because they

    A. facilitate nursing intervention.
    B. give direction for dependent nursing actions.
    C. explain the medical diagnoses.
    D. Define the focus of nursing to others.

37. A weakness in the present nursing diagnoses taxonomy is a lack of

    A. direction for future development.
    B. psychiatric diagnoses.
    C. health promotion diagnoses.
    D. medical etiologies.

38. Choose the diagnostic statement that follows the PES format.

    A. alteration in comfort: pain, related to swelling in ankle manifested by the inability to walk.
    B. congested lungs due to pneumonia resulting in coughing and elevated temperature.
    C. lack of appetite related to anorexia causing weight loss.
    D. cirrhosis of the liver evidenced by increase in abdominal fluid, jaundice, and hemorrhage.

39. The movement for the development of nursing diagnoses

    A. was initiated by Gordon and Kim.
    B. was first sponsored by MissouriUniversity School of Nursing.
    C. is recognized as the American Nursing Diagnosis Consortium.
    D. Was initiated by Gebbie and Lavin.

ADDITIONAL LEARNING ACTIVITIES

1. Review the client records of your assigned client. Determine what information was used to arrive at the nursing diagnoses for your client.

2. Make a list of the nursing diagnoses you find for your client. Use the guidelines given in Chapter 11 of the textbook to determine whether they were written correctly.

3. Interview a nurse on your clinical unit to find out how using nursing diagnoses improves his or her efficiency in providing quality nursing care.

4. Review the most recent list of approved nursing diagnoses approved by NANDA. How are the diagnoses grouped?

12 PLANNING

Planning is the most crucial phase of the nursing process since all the nursing interventions that will follow are decided on at this time. Intelligent and careful planning prevents, reduces, or eliminates the health problems for the client and increases the efficiency of the care-givers. Since the client is the focus of nursing care, it is imperative that he or she be included in the decision-making process, where health problems are identified and prioritized, and outcome criteria are formulated. At the completion of this chapter the student will be able to:

- Describe the essential aspects of the planning phase of the nursing process.
- Identify four components of planning.
- Identify criteria that help the nurse and client set priorities.
- State the purposes of establishing client goals.
- Describe the relationship of goals to the nursing diagnoses.
- Differentiate between goals and outcome criteria.
- Identify guidelines for writing goals and outcome criteria.
- Describe three aspects of planning nursing strategies.
- Identify major purposes of a written care plan.
- Identify various formats used for nursing care plans
- Identify essential guidelines for writing nursing care plans.
- Describe the consulting process.
- Describe effective approaches to discharge planning.

DEFINITION AND PROCESS

Complete the exercise about the planning phase by filling in the blanks.

The planning phase of the nursing process is defined as 1 ________________

The following persons can be involved in planning, the 2 ________________,

3 ________________, 4 ________________ and 5 ________________. The

planning process uses 6 ____________________ obtained during assessing and

7 ____________________ that present the client's health problems.

COMPONENTS OF PLANNING

Plans Bingo: Identify the term described below, then fill in each box in the game with the appropriate term

	P	L	A	N	S
1	8.	13.	18.	23.	28.
2	9	14.	19.	24.	29.
3	10.	15.	20.	25.	30.
4	11.	16.	21.	26.	31.
5	12.	17.	22.	27.	32.

P 8 Self-care deficit
P 9 Establishing an order for nursing strategies
P 10 Broad statement about the expected change in a client
P 11 Model used for judging
9 12 Designing nursing strategies for the client
L 13 Inferring facts from known data
L 14 Actions the nurse takes to assist client
L 15 Activities that maintain or restore the client's usual pattern
L 16 Priority framework
L 17 Synonymous with outcome criteria
A 18 Describe specific responses of the client
A 19 Desired outcome in client behavior
A20 Actions chosen to treat a specific nursing diagnosis
A 21 Technique used by a group to develop solutions
A 22 Written guide that organizes information about a client's health
N 23 Scientific reason for selecting a nursing action
N 24 File of specially designed nursing care plan cards
N 25 Predicting which actions will solve a problem
N 26 Includes date, action verb, content, time, and signature

N 27 Broadly stated client outcome
S 28 Provide a direction and time span for planned activities
S 29 Include subject, verb, condition modifiers, and criterion for desired performances
S 30 Are guides for assigning staff and documenting care
S 31 Must be within the standards of care determined by state law
S 32 Shows the nurse's accountability

TRUE OR FALSE

33. _______ A major purpose for establishing client goals is to provide direction for planning nursing interventions.

34. _______ Client goals are derived from the etiology clause of the nursing diagnosis.

35. _______ The written plan of care gives direction for nursing care, provides continuity of care, and serves as a guide for assigning staff.

36. _______ Most plans of care are organized into four columns; nursing diagnosis, goals, nursing strategies, and outcome criteria.

37. _______ Nursing care plans should include complete terminology rather than abbreviations that could be misunderstood.

38. _______ Nurses use the consulting process to verify findings, implement change, and obtain additional knowledge.

Identify the incorrect outcome criteria with a checkmark

39. _______ Move the client from side to side.

40. _______ The client will walk for longer periods of time.

41. _______ the client will feel less pain within 1 hour.

42. _______ The client will drink 1 glass of water every hour.

43. _______ The nurse will discuss the effects of smoking with the client.

44. _______ The client will perform range-of-motion exercises on his right side three times a day.

Develop at least two outcome criteria for each of the nursing diagnoses that follow.

Bathing self-care deficit related to neurological impairment

45. ____

46. ____

Activity intolerance related shortness of breath

47. ____

48. ____

DISCHARGE PLANNING

Describe at least two approaches the nurse must take in order to establish effective discharge planning for the client.

49. ____

____

50. ____

____

SELF ASSESSMENT QUESTIONS

51. Which of the following problems should be given the highest priority using Maslow's hierarchy of needs?

    A. Sexual dysfunction
    B. Anger
    C. Social isolation
    D. Malnutrition

52. An example of subjective data is

    A. pain.
    B. edema.
    C. pulse.
    d. vomiting.

53. Which of the following outcome criteria is stated correctly?

    A. Drinks eight 6-ounce glasses of water between 8 AM and 8 PM
    B. Walks 3 times a day
    C. Includes more activity in daily routine each day
    D. Discusses proper administration of medication with nurse

54. Outcome criteria describe the

    A. nurse's objectives.
    B. client's responses.
    C. long-term goals.
    D. short-term strategies.

55. In the planning phase of the nursing process the nurse

    A. collects client data.
    B. evaluates nursing care.
    C. predicts client responses.
    D. diagnoses client problems.

ADDITIONAL LEARNING ACTIVITIES

1. Review the nursing diagnosis that you developed for your assigned client in the clinical unit. Develop at least two outcome criteria for each diagnosis.

2. Review a chart of a client on the clinical unit. Make a list of the outcome criteria that have been developed for the client. Use the guidelines give in Dhapter 12 to determine whether they were written correctly.

3. Interview a nurse who determines the effectiveness of nursing care by evaluating to what degree outcome criteria have been met. Find out how the nurse uses this information to modify nursing care.

4. Write a care plan for an assigned client including an assessment, diagnoses, outcome criteria, and nursing actions. Include rationales for your nursing actions.

13 IMPLEMENTING

Implementing, the third phase of the nursing process, requires that the nurse use independent, dependent, and collaborative nursing actions in providing nursing care. The nurse uses cognitive, interpersonal, and technical skills to collaborate with the client whenever possible about the nursing care that is to be given. On completing this chapter the student will be able to:

- Differentiate independent, dependent, and collaborative nursing actions.
- Compare protocols and standing orders.
- Identify five essential aspects of implementing nursing strategies.
- Describe two processes that continuously operate throughout the implementing phase.
- Describe three categories of skills used to implement nursing strategies
- Identify four cognitive skills nurses use when implementing nursing strategies.
- Differentiate critical and creative thinking.
- Identify essential guidelines for implementing nursing strategies.

IMPLEMENTING DEFINED AND TYPES OF NURSING ACTIONS

Define the following terms:

1. Implementing ______________________________

____

2. Independent nursing actions ______________________________

____

3. Accountable ______________________________

____

4. Taxonomy ______________________________

____

5. Dependent nursing actions ____

____

6. Collaborative nursing actions____

____

7. Protocol ____

____

8. Standing order ____

____

PROCESS OF IMPLEMENTING

List the five essential aspects of the implementing phase.

9. ____

10. ____

11. ____

12. ____

13. ____

Discuss Kozier and Erb's statement, "Reassessing the client and validating the nursing care plan are subprocesses that operate continuously throughout the implementing phase."

14. ____

____

____

____

IMPLEMENTING SKILLS

Identify and describe the three categories of skills needed to implement nursing actions in the diagram below.

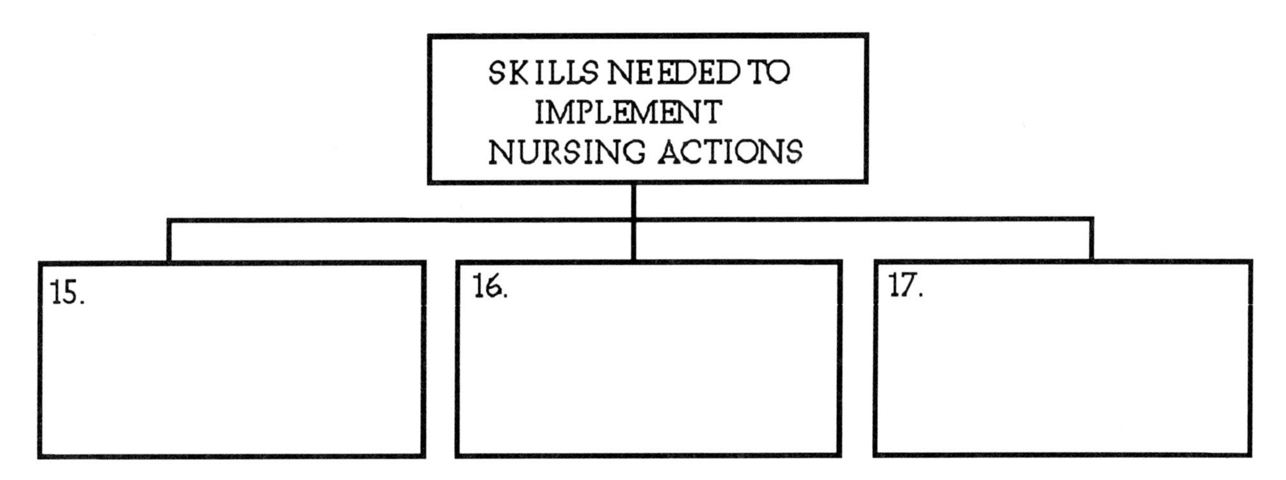

Identify the types of thinking described below.

18. ____________________ has a goal and is purposeful. A person uses this type of thinking when trying to form a judgment.

19. ____________________ has very little direction and often involves random thoughts.

20. ____________________ is a pattern of thinking based on knowledge, experience and the ability to conceptualize and analyze relationships.

21. ____________________ An example is daydreaming.

22. ____________________ After assessing a client's adverse reaction to a drug, a nurse decides not to administer the next dose.

23. ____________________ is establishing relationships and new concepts in solving problems innovatively.

IMPLEMENTING ACTIVITIES

Briefly describe the eight guidelines Kozier and Erb recommend for implementing nursing strategies.

24. ______________________________

25. ______________________________

26. ______________________________

27. ______________________________

28. ______________________________

29. ______________________________

30. ______________________________

31. ______________________________

SELF ASSESSMENT QUESTIONS

32. Dependent nursing actions are

    A. activities performed jointly with another member of the health team.
    B. activites carried out on the order of a physician.
    C. activities that the nurse initiates as a result of the nurse's own knowledge and skill.
    D. activities carried out during the physical examination.

33. The difference between a protocol and standing orders are

    A. protocols are written about procedures and standing orders are written about policies.
    B. protocols are verbal orders , standing orders are written.
    C. protocols govern independent nursing actions, standing orders dictate dependent nursing actions.
    D. protocols are collaborative efforts between the nurse and the physician whereas standing orders are dictated by the physician.

34. The pattern of thinking based on knowledge, experience and the ability to conceptualize and analyze relationships is called

    A. associative thinking.
    B. directed thinking.
    C. creative thinking.
    D. critical thinking.

35. Cognitive skills include

    A. problem solving.
    B. decision making.
    C. creativity.
    D. all of the above.
    E. none of the above.

36. Which of the following statements is not true? Nursing actions

    A. are based on scientific knowledge.
    B. may require teaching, and supportive measures.
    C. should respect the dignity of the client.
    D. are based on physician's orders.

ADDITIONAL LEARNING ACTIVITIES

1. Review a chart of a client on the clinical unit. Make a list of the nursing strategies that have been developed for the client. Use the guidelines given in Chapter 13 to determine whether the nursing strategies were written correctly.

2. Review the outcome criteria that you developed for your assigned client. Develop at least two nursing strategies for each outcome criteria.

3. Write a care plan for an assigned client including an assessment, diagnoses, outcome criteria and nursing actions. Implement the plan of care for the client using the skills outlined in Chapter 13.

4. Observe a nurse providing primary nursing care. What independent nursing actions were carried out? What dependent actions were carried out?

EVALUATING

Evaluating the fifth and last phase of the nursing process, identifies to what degree the client's goals have been met. Like the other phases of the nursing process, evaluating enhances the quality of nursing care by measuring its effectiveness against outcome criteria that have been previously established by and for the client. Evaluating ensures that client goals are met, nursing interventions are effective, and client problems are reduced or eliminated. After completing this chapter the student will be able to:

- Describe six components of the evaluation process.
- Describe the steps involved in reexamining the client's care plan both when goals are met and when they are not met.
- Differentiate quality assessment from quality assurance.
- Describe three approaches to quality evaluation.
- Identify essential steps in developing tools to evaluate quality care.
- Identify various methods used to evaluate nursing care.

DEFINITION OF EVALUATION

In the space below, write a definition of the evaluation process in your own words.

1. ____

____

Briefly explain the following statements:

Evaluating is both a concurrent and a terminal process.

2. ____

____

Evaluating is a purposeful and organized activity.

3. ______________________________

____

PROCESS OF EVALUATING

Identify and describe each of the six components in the evaluation process.

4. ______________________________

5. ______________________________

6. ______________________________

7. ______________________________

8. ______________________________

9. ______________________________

After experiencing pain in his right flank area, Dan Matthews was admitted to the hospital. His short-term goals were:

1. Drink eight glasses of water between 7 AM and 7 PM daily.
2. Consume a 1200 calorie diet daily.
3. Walk the length of the hospital corridor three times a day.

Dan was only able to drink four glasses of water a day due to persistent nausea and vomiting. He was unable to eat any solid food. He did, however, walk the length of the hall three times a day. Describe whether the three goals stated above were met or unmet.

10. Goal 1 ____________

11. Goal 2 ____________

12. Goal 3 ____________

Describe the nurse's responsibility when a goal is not met.

13. ____

____

EVALUATE THE QUALITY OF NURSING CARE

Define quality assessment, quality assurance, and the difference between the two terms in the appropriate boxes.

14. Quality Assessment

15. Quality Assurance

16. Difference

Describe the steps used in developing tools for measuring quality of care.

17.

18.

19.

20.

SELF ASSESSMENT QUESTIONS

21. Process standards that provide the profession with a framework for delivery and evaluation of care are published in the

    A. nurse practice acts of each state.
    B. NLN's *Evaluation and Nursing Practice.*
    C. ANA's *Standards of Practice.*
    D. NANDA's *Diagnostic Manual.*

22. Assessment tools that are used to measure quality of nursing care

    A. consist of standards and criteria.
    B. measure process and outcome.
    C. use similar scoring systems.
    D. use similar methods.

23. A PRO or professional review organization

    A. provides utilization reviews.
    B. develops standards of care.
    C. monitors quality and cost of care.
    D. carries out all of the above functions.

24. An encounter between two persons equal in education, abilities, and qualifications during which one person critically reviews the practices that the other has documented in a client's record is called a

    A. nursing review.
    B. peer review.
    C. nursing audit.
    D. retrospective audit.

25. A hospital is not legally required to contract with a Professional Review Organization unless JCAH has determined that the quality of care has been neglected. This statement is

    A. true.
    B. false.

ADDITIONAL LEARNING ACTIVITIES

1. Review the outcome criteria that you developed for your assigned client on the clinical unit. After caring for the client evaluate the effectiveness of your nursing actions by classifying the outcome criteria as being met, partially met or not met.

2. Review a chart of a client who has been hospitalized for at least one week. Determine how goals were evaluated. How did the evaluative statements impact or modify nursing care?

3. Discuss how peer review is usd in your clinical area with a professional nurse. Is she or he aware of the process that is being used. Find out how peer review affects nursing care in this agency.

15 HELPING AND COMMUNICATING

Throughout the entire nursing process the nurse uses communication skills to elicit data, disseminate information, teach concepts, and demonstrate skills to individuals and groups. This chapter focuses on the communication and interpersonal knowledge and skills the nurse must possess to work with families and groups. After completing this chapter the student will be able to:

- Describe four phases of the helping relationship.
- Describe essential aspects of communication and the communication process.
- Explain the four elements of the communication process outlined in this chapter.
- Identify ways in which selected factors influence the communication process.
- Differentiate verbal and nonverbal communications.
- Give guidelines for assessing problems in communicating.
- List examples of nursing diagnoses pertaining to communication.
- Describe strategies for planning to resolve communication problems.
- Describe effective and ineffective methods used by nurses when communicating with clients.
- List outcome criteria that can be used to evaluate whether or not communication problems have been resolved.
- Identify features of effective groups.

THE HELPING RELATIONSHIP

Describe the tasks and skills associated with each step of the helping relationship.

PHASE	TASKS	SKILLS
Preinteraction	1.	5.
Introductory	2.	6.
Working	3.	7.
Termination	4.	8.

COMMUNICATION IN NURSING

Match the following terms in column I with the appropriate description in column II.

9. _______ Communication

10. _______ Social communication

11. _______ Structured communication

12. _______ Therapeutic communication

13. _______ Verbal communication

14. _______ Nonverbal communication

A. Unplanned communication carried out in an informal setting

B. Uses the spoken or written word to communicate.

C. A process that helps overcome temporary stress, to get aong with other people, to adjust to the unalterable, and to overcome psychological blocks.

D. The interchange of information between two or more people

E. Uses various forms of communication such as gestures or facial expressions

F. Definite planned content

Differentiate between verbal and nonverbal communication. Give an example of each.

15. ____

____

List five criteria the nurse needs to consider when communicating with clients.

16. ____

17. ____

18. ____

19. ____

20. ______________________________

THE COMMUNICATION PROCESS

Draw a diagram depicting the communication process in the box below.

Describe intimate, personal, social and public space in the space below.

21. Intimate space ______________________________

____

22. Personal space ______________________________

____

23. Social space ______________________________

____

24. Public space ______________________________

____

LANGUAGE DEVELOPMENT

Describe the various stages of language development in the appropriate space.

25. Babbling ______________________________

____

26. Echolalia ____

____

27. Holophrastic speech ____

____

28. Egocentric speech ____

____

29. Socialized speech ____

____

ASSESSING COMMUNICATION

Briefly describe the factors the nurse must consider when assessing communication.

LANGUAGE DEVELOPMENT	NONVERBAL BEHAVIOR	STYLE OF COMMUNICATION
30.	35.	40.
31.	36.	41.
32.	37.	42.
33.	38.	43.
34.	39.	44.

DIAGNOSING COMMUNICATION PROBLEMS

Case Study

"I won't go in there again! He threw his coffee pot at me. I don't have to take that kind of treatment." Denise Mackey, RN, was upset. She had just come from Mr Jarguzis' room for the fifth time. "I can't please him no matter what I do," she complained to the head nurse. "When I asked him what he needed, he got mad and said that I was the expert and I should know what he needed." Mr Jarguzis, a 68-year-old retired carpenter, had been admitted the previous day with a medical diagnosis of a duodenal ulcer. On admission he seemed very anxious and frightened. He refused to answer the admitting nurse's questions and insisted on going home. Beatrice Romano RN, the unit manager, told Denise, "I think Mr Jarguzis is worried and frightened. We're going to have to help him deal with his fears."

Develop a nursing diagnosis for Mr. Jarguzis based on the information given.

45. ____

____

PLANNING FOR EFFECTIVE COMMUNICATION

Develop three nursing interventions for Mr Jarguzis that would address his communication problems.

46. ____

47. ____

48. ____

Develop three outcome criteria for Mr Jarguzis.

49. ____

50. ____

51. ____

IMPLEMENTING AND EVALUATING COMMUNICATION

Denise went in to talk to Mr Jarguzia later in the morning, after she had calmed down. A process recording of their conversation follows. Identify the technique that Denise used in the communication technique column. Indicate if the techique was positive with a (+), or if it was negative with a (-).

Denise, RN	Mr Jarguzis	Technique (+) or (-)
52. "Good morning, Mr Jarguzis. I'm going to be caring for you the next few mornings. My name is Denise Martin." (Sits down 2 feet from client's bed.)	"Hello." (Looks suspiciously at nurse.)	
53. "What brings you to the hospital?" (Crosses legs, leans back in chair.) "I need to know if I'm going to care for you." (Begins to shake right foot.)	"I've got a hernia. What's it to you? Care! What's that? Nobody around here cares for any body!"	
54. "I'm not sure I know what you mean." (Uncrosses leg, leans toward client.) "You seem angry. Has someone offended you?"	" Nobody cares! Is that clear enough? If you call inefficiency and ignorance offending me, yes!"	
55. "Has something happened to upset you?"	"Yes! Do you really want to hear this?"	
56. "Yes I do. I want to help." (Looks client squarely in the eye.)	"It's such a little thing, It's almost stupid."	
57. "A little thing, you say. You feel it's stupid."	"Yes. Getting so upset over a pot of coffee is stupid."	

Denise, RN	Mr Jarguzis	Technique (+) or (-)
58. "Well, the coffee must be important to you."	"Well, you see, I have this problem. I get so constipated that if I drink a hot cup of coffe every morning at 6:00 AM. It gets my insides moving."	
59. "A cup of hot coffee at 6:00 AM every morning prevents constipation for you?"	"You got it! I need hot coffee by 6:00. If I don't get it, I ge constipated. But my breakfast doesn't get here until 9:00 AM on this floor. I can't make those fools out there understand."	
60. "Don't call the staff fools, Mr Jarguzis. They're very good people. They just don't understand your problem." (Crosses arms over chest.)	"If they can't understand a simple problem like this, how can I depend on them if I have to have surgery?"	
61. "I think you're drawing wrong conclusions. They're excellent nurses." (Looks at clock on the wall.)	"That remains to be seen!"	
62. "Well, Mr Jarguzis, I think we can solve this coffee problem. I'll tell the staff you need a pot of hot coffee every morning at 6:00 AM. We'll also check you daily to make sure you're not becoming constipated." (Gets up, and touches client's arm.)	"Thanks for listening."	

GROUP INTERACTION

List the three main functions of an effective group.

63. ______________________________

64. ______________________________

65. ______________________________

SELF ASSESSMENT QUESTIONS

66. Which of the following comments would be most helpful in establishing a nurse/ client relationship with Mr. Jarguzis?

    A. "I'll be planning your care for the next several days."
    B. "How do you feel about being in the hospital?"
    C. "I'll do everything I can to please you."
    D. "I've cared for many clients with problems like yours."

67. The day before Mr Jarguzis is scheduled for surgery he yells at the nurse, "Get out of my room!" The nurse's most therapeutic response might be

    A. "You're upset. I'll come back later."
    B. "Why are you acting this way?
    C. "I'll find another nurse to take care of you."
    D. "I'm sorry you feel this way."

68. The characteristic of the nurse that would most allay Mr Jarguzis's anxiety is

    A. empathy.
    B. kindness.
    C. reliability.
    D. warmth.

69. Carmella Egizio is the vice president of a large advertising agency on the east coast. She chairs the research committee whose members determine marketing strategies for the coming year. This group can be categorized as

    A. primary.
    B. secondary.
    C. convenience.
    D. self-protective.

70. George Lee has organized a group composed of parents like himself who have lost a child through death. He hopes the group will help the parents cope with the loss of their children. This is an example of a

    A. task group.
    B. self-help group.
    C. teaching group.
    D. brain-storming group.

ADDITIONAL LEARNING ACTIVITIES

1. Make a tape recording of a group meeting with the permission of the members. Identify the role each member of the group plays.

2. Draw a sociogram of the group selected in activity 1. Analyze the strengths and weaknesses in the interaction patterns in the group. Discuss ways these weaknesses could be alleviated.

3. Interview a client in a clinical setting. Write a process recording of your interactions. Evaluate the effectiveness of your communications skills.

16 TEACHING, LEARNING, AND PLANNED CHANGE

In today's health care delivery system the spiralling costs of health care and the initiation of the prospective payment system have caused earlier and earlier discharge of clients from the hospital. Because of this trend in early discharge, the client is often expected to learn to provide care for him- or herself at home that was formally provided by a nurse in a hospital setting. As a result, the methods the nurse uses, and the ways in which the client learns have become an integral part of the nursing process. This chapter focuses on teaching and learning methods as well as the process of planned change. After completing this chapter the student will be able to:

- Explain how andragogy can guide client teaching.
- Describe types of change.
- Explain three theories of change.
- List factors that facilitate learning.
- Explain the three domains of learning.
- Outline five principles of teaching.
- Describe essential aspects of a teaching plan.
- Explain essential factors in assessing for learning.
- Identify eight guidelines that help plan teaching.

FACILITATING LEARNING

Match the terms in column I with the appropriate descriptions in column II.

1. _______ Learning need
2. _______ Learning
3. _______ Behaviorism
4. _______ Cognitivism
5. _______ Angragogy
6. _______ Pedagogy
7. _______ Developmental change
8. _______ Covert change

A. Learning is a cognitive activity
B. The physiopsychosocial changes that occur during the life cycle
C. The transfer of knowledge could occur if the new situation resembles the old situation
D. Helping children learn
E. A need to change behavior
F. Without the person's awareness
G. A change in behavior
H. Helping adults learn

PRINCIPLES OF LEARNING

In the table that follows list and briefy describe three factors that facilitate learning and three factors that inhibit learning.

FACTORS THAT FACILITATE LEARNING	FACTORS THAT INHIBIT LEARNING
9.	12.
10.	13.
11.	14.

List and briefly describe Bloom's three domains of learning.

15. ______________________________

16. ______________________________

17. ______________________________

TEACHING

List the five principles of learning.

18. ______________________________

19. ______________________________

20. ______________________________

21. ______________________________

22. ______________________________

ASSESSING, DIAGNOSING, PLANNING, AND IMPLEMENTING

Case Study

Bob Jenkins is 52-year-old high school teacher who was hospitalized three days ago with complaints of severe pain that radiates from the lumbar spine down through his left leg. Bob's condition has been diagnosed as a ruptured intervertebral disc which is aggravated by obesity. His doctor has prescribed a 1200 - calorie weight reduction diet. When Bob's nurse approaches him with information about the diet, he says, "I don't need to read this stuff. I'll just cut down on a couple of beers a day. That should do it."

Bob does not seem ready or motivated to learn about his diet. What can the nurse do to stimulate these factors when implementing his plan of care?

23. ______________________________

____

____

After reading the information about Bob, develop a nursing diagnosis that reflects his probblem related to learning.

24. ______________________________

____

Develop one long-term goal and one short-term objective for Bob.

25. Long-term goal ______________________________

26. Short-term objective ______________________________

Describe two teaching strategies that will be useful in helping Bob learn the 1200 calorie diet.

27. ______________________________

28. ______________________________

List at five guidelines the nurse can use to implement Bob's teaching plan.

29. ______________________________

30. ______________________________

31. ______________________________

32. ______________________________

33. ______________________________

SELF ASSESSMENT QUESTIONS

34. A client's ability to learn would most likely be facilitated by his?

    A. readiness.
    B. anxiety.
    C. illness.
    D. I.Q.

35. One of the best ways to increase a client's motivation to learn is to

    A. explain the need to learn.
    B. create a comfortable environment.
    C. provide constructive feedback.
    D. use a variety of visual aids.

36. Affective learning involves

    A. knowledge.
    B. attitudes.
    C. skills.
    D. culture.

37. One of the first steps in developing a teaching plan is to

    A. develop evaluative criteria.
    B. determine teaching priorities.
    C. set up a class schedule.
    D. document the correct information.

38. The three types of information found in learning objectives should be

    A. positive feedback, encouragement, and motivation.
    B. long-range outcome, behavior, and data.
    C. learning activity, nurse behavior, and long-term need.
    D. performance, conditions, and criteria.

ADDITIONAL LEARNING ACTIVITIES

1. Assess the learning needs and develop a teaching plan for one of your clients in a clinical setting.

2. Attend an education class for new diabetics in your clinical setting conducted by a nurse educator. How does the nurse educator implement the prinicples of learning and teaching in his or her class?

3. Present a teaching plan that you have developed for a client to your class. Ask the class to critique your presentation.

17 DOCUMENTING AND REPORTING

The major reason for developing communication skills in nursing is to provide coordinated, high-quality care for clients. When communication is on target and focused, all members of the health team who are responsible for a client's care are well informed about the client needs and duplication of nursing activities is avoided. The best method of ensuring efficient communication is through accurate written transmission of information. This chapter focuses on the knowledge and skills the nurse needs to develop and implement efficient communication. After completing this chapter the student will be able to:

- Identify seven purposes of client records.
- Describe the components of source-oriented medical records and problem-oriented medical records (POMR).
- Describe three types of progress records.
- Differentiate between narrative, SOAP, focus charting, and charting by exception.
- Identify measures used to ensure that recording meets legal standards.
- Identify abbreviations and symbols commonly used for charting.
- Identify measures used to maintain the confidentiality of client records.
- Describe the change-of-shift report.
- Identify essential guidelines for reporting client data.
- Compare the advantages and disadvantages of nursing care conferences and nursing care rounds.

COMMUNICATION AMONG HEALTH TEAM MEMBERS

State four reasons written and verbal communication in nursing must be accurate and complete.

1. ______________________________

2. ______________________________

3. ______________________________

4. ______________________________

PURPOSE OF CLIENT RECORDS

List seven purposes for maintaining client records.

5. ______________________________
6. ______________________________
7. ______________________________
8. ______________________________
9. ______________________________
10. ______________________________
11. ______________________________

TYPES OF RECORDS

List five components of the source-oriented medical record.

12. ______________________________
13 ______________________________
14. ______________________________
15. ______________________________
16. ______________________________

List four components of the problem-oriented medical record.

17. ______________________________
18. ______________________________
19. ______________________________
20. ______________________________

Briefly describe the various types of records listed below.

21. Source-oriented medical records ______________________________

____

22. Problem-oriented medical records ______________________________

____

23. Kardex ______________________________

____

24. Computer records ____

____

TYPES OF PROGRESS RECORDS

Describe each of the following types of progress records.

25. Nurse's notes ____

____

26. Flowsheets ____

____

27. Discharge notes ____

____

FORMATS FOR WRITING PROGRESS NOTES

Identify the method for writing progress notes by matching the method in column I with its description in column II.

28. ________ Narrative charting

29. ________ SOAP format

30. ________ Focus charting

31. ________ Charting by exception

A. Uses key words that describe what is happening to the client.

B. A documentation system in which only significant findings are recorded.

C. A chronologic description of information.

D. An acronym for subjective data, objective data, assessment and planning.

GUIDELINES ABOUT RECORDING

32. Correct mistakes in the following charting.

Date	Time	Notes
10/91	6:00	Complains of pain in left lower lag. Left leg elevated on 2 pillows. *J. Burns R.N.*
		Vomited large amount of green mucus tinged liquid with after eating breakfast.

Identify the following abbreviations and symbols used in charting.

33. abd ______________________
34. c/o ______________________
35. gtt ______________________
36. hs ______________________
37. OOB ______________________
38. Tr. ______________________
39. ♀ ______________________
40. ♂ ______________________
41. X ______________________
42. # ______________________
43. < ______________________

REPORTING

The following news item appeared in the paper about a hospitalized client.

> Maggie Mitchell, a 64-year-old woman, was hit by a car in the downtown metro area today. Witnesses say she appeared to be hallucinating as she walked into oncoming traffic. Ms Mitchell, a bag lady, has been a familiar figure in the downtown area since she was released from Montgomery Psychiatric Hospital five years ago. She is in fair condition in County General Hospital where laboratory tests have uncovered elevated blood sugar and alcohol levels. Hospital personnel have indicated that these findings may have led to the accident.

Was Ms Mitchell's right to privacy invaded? How?

44. ____

____

____

CONFERRING

Describe the advantages of each of the following.

45. Nursing care conference ______________________________

____

46. Nursing care rounds ______________________________

____

SELF ASSESSMENT QUESTIONS

47. The major disadvantage of narrative charting is that it

    A. is not written in a chronological manner.
    B. records both dependent and independent nursing interventions.
    C. uses infrequently used symbols.
    D. records data about a specific problem in several places.

48. A change of shift report is

    A. an oral report given face-to-face.
    B. given from one RN to another.
    C. is given by audiotape.
    D. all of the above.
    E. none of the above.

49. To consult another person or persons for advice, information, ideas, or instructions is

    A. communicating.
    B. conferring.
    C. consulting.
    D. consorting.

50. All of the following may have access to a client's legal records without the client's consent *except*

    A. a health care provider involved in the client's care.
    B. an insurance company responsible for paying a client's bill.
    C. student and graduate nurses for use in cllient conferences.
    D. the attending physician.

51. Which of the following guidelines should *not* be included in a change-of-shift report?

    A. Identify the client by name, room number and bed designation.
    B. Note any significant changes in the client's condition.
    C. Include normal as well as abnormal information.
    D. Provide exact information.

ADDITIONAL LEARNING ACTIVITIES

1. Collect forms used for documenting and reporting data from a variety of clinical settings. Discuss the advantages and disadvantages of each type.

2. Attend a change-of-shift report and/or nursing rounds. Discuss the advantages and disadvantages of each type.

18 ASSESSING VITAL SIGNS

Assessing vital signs is an important aspect of the overall assessment the nurse performs on the client in the initial phase of the nursing process. Vital signs are important because they reflect significant changes in body function that may be otherwise overlooked. This chapter focuses on assessment of the temperature, pulse, respirations, and blood pressure. After completing this chapter the student will be able to:

- Define terms and abbreviations used when measuring body temperature, pulse, respirations, and blood pressure.
- Describe five factors influencing the body's heat production.
- Identify four ways in which the body loses heat.
- Describe the body's temperature regulating system.
- Compare oral, rectal, and axillary methods of measuring body temperature.
- Identify situations in which specific methods for measuring body temperature are indicated or contraindicated.
- Identify recommended intervals required to obtain accurate temperature readings for each method and for different types of equipment.
- Describe selected alterations of body temperature and appropriate nursing care for these alterations.
- Identify nine pulse sites commonly used to assess the pulse and state the reasons for their use.
- List the characteristics that should be included when assessing pulses.
- Explain how to measure the apical pulse and apical-radial pulse.
- Describe the mechanics of breathing and the mechanisms that control respirations.
- Identify the characteristics that should be included in a respiratory assessment.
- Differentiate systolic from diastolic blood pressure.
- Describe various methods and sites used to measure blood pressure.

VITAL SIGNS

Complete the following paragraph by filling in the blanks.

The vital or 1 ______________ signs are usd to monitor the 2 ______________ of the

body. The signs should be evaluated with reference to the client's present and prior

3 ______________________. Vital signs include the 4 ______________________

5 ______________________, 6 ______________________ and 7 ______________________.

BODY TEMPERATURE

Briefly describe the five factors that affect the body's heat production in the diagram.

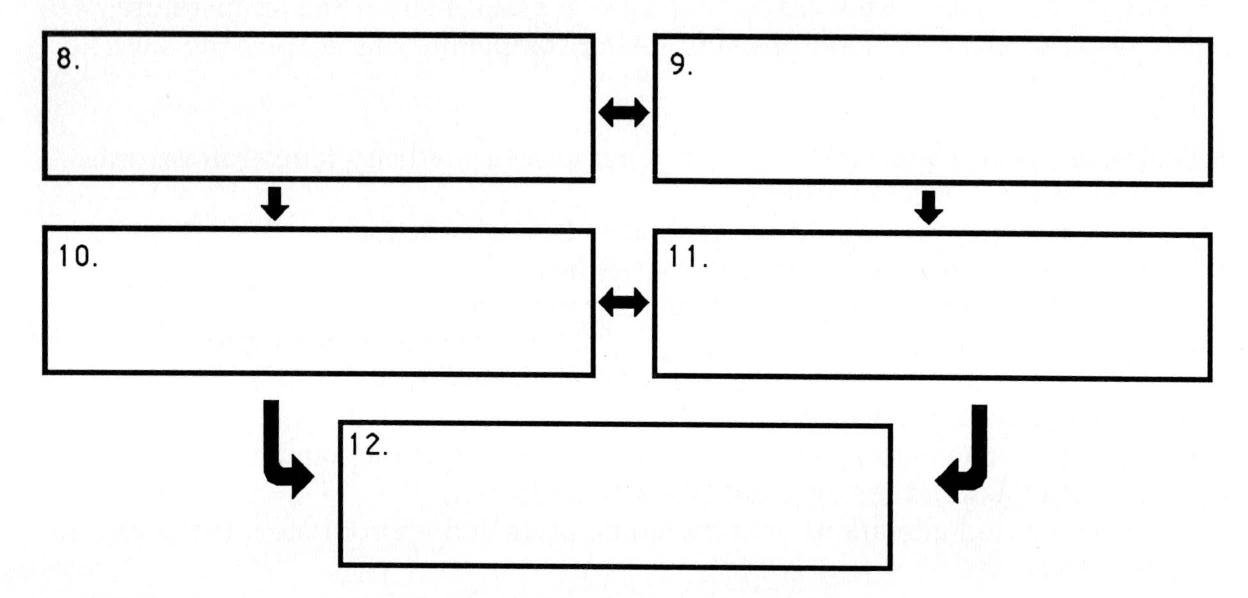

Identify the terms in column I by placing the appropriate letter from column II in the space provided.

13. ______ Conduction
14. ______ Convection
15. ______ Vaporization
16. ______ Hypothalamic integrator
17. ______ Constant fever
18. ______ Crisis
19. ______ Hypothermia

A. Transfer of heat between molecules
B. Another term for evaporation
C. Controls the core temperature
D. Temperature fluctuates minimally but remains elevated
E. Dispersion of heat by air currents
F. Core body temperature below the lower limit of normal
G. Also called a flush

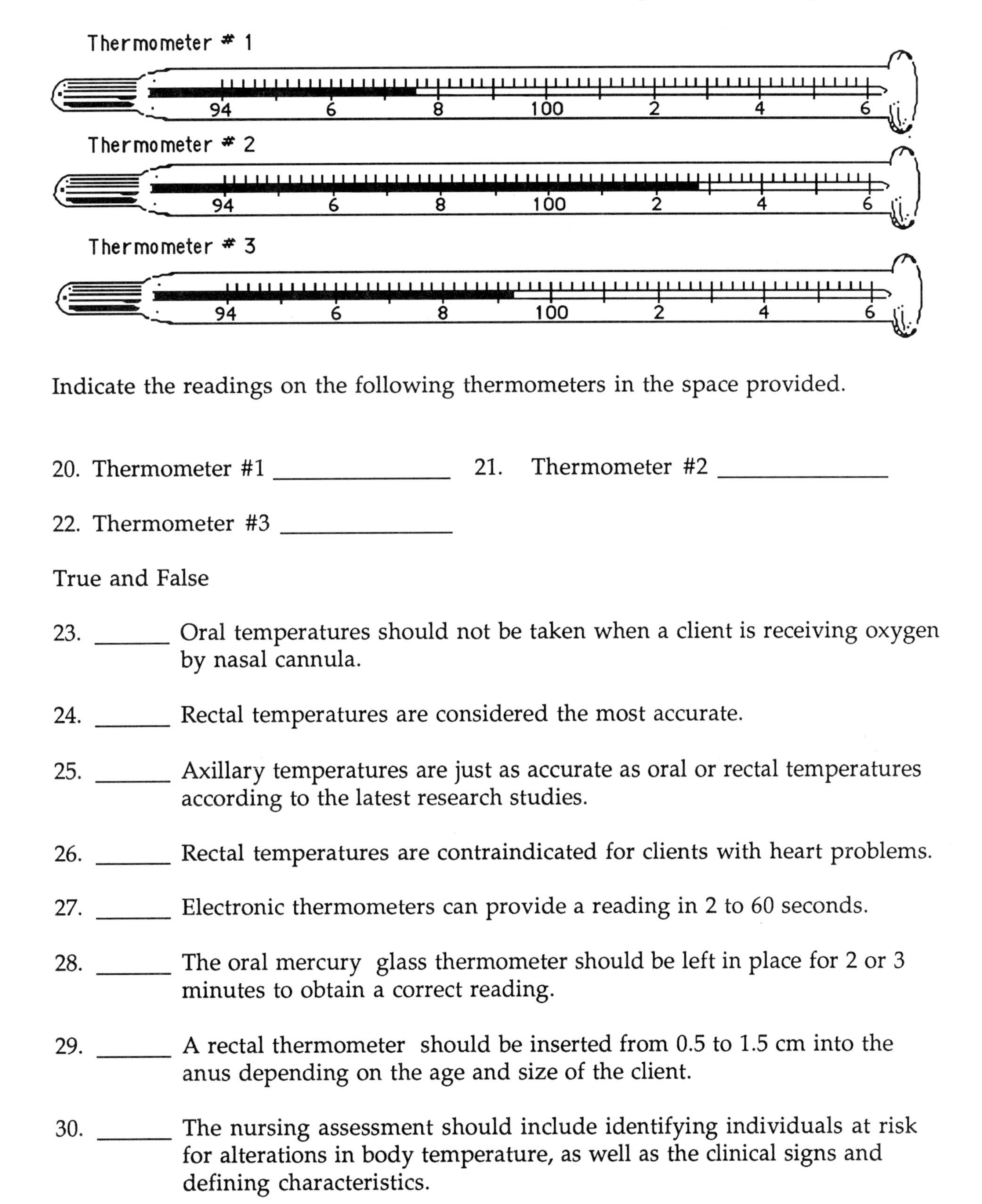

Indicate the readings on the following thermometers in the space provided.

20. Thermometer #1 ______________ 21. Thermometer #2 ______________

22. Thermometer #3 ______________

True and False

23. ______ Oral temperatures should not be taken when a client is receiving oxygen by nasal cannula.

24. ______ Rectal temperatures are considered the most accurate.

25. ______ Axillary temperatures are just as accurate as oral or rectal temperatures according to the latest research studies.

26. ______ Rectal temperatures are contraindicated for clients with heart problems.

27. ______ Electronic thermometers can provide a reading in 2 to 60 seconds.

28. ______ The oral mercury glass thermometer should be left in place for 2 or 3 minutes to obtain a correct reading.

29. ______ A rectal thermometer should be inserted from 0.5 to 1.5 cm into the anus depending on the age and size of the client.

30. ______ The nursing assessment should include identifying individuals at risk for alterations in body temperature, as well as the clinical signs and defining characteristics.

31. ______ Oral temperatures can be measured less quickly than rectal temperatures.

PULSE

In the diagram below mark the sites of the radial, brachial, dorsalis pedis, posterior tibial, carotid, and femoral pulses with an X.

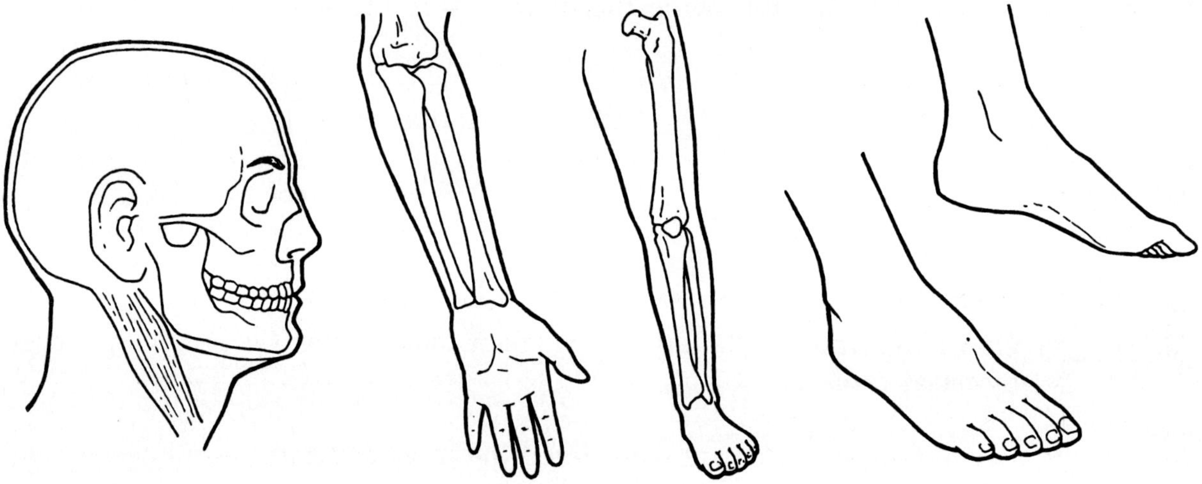

Describe the location where the apical pulse is auscultated in the adult. How is the apical pulse measured.

32. ____

____

____

List five characteristics the nurse observes when assessing the pulse.

33. ____

34. ____

35. ______________________________

36. ______________________________

37. ______________________________

RESPIRATIONS

Describe the mechanics of inhalation and exhalation.

38. ______________________________

____

____

____

List four characteristics the nurse observes when assessing respirations.

39. ______________________________

40. ______________________________

41. ______________________________

42. ______________________________

BLOOD PRESSURE

What is the difference between systolic and diastolic blood pressure?

43. ______________________________

____

____

SELF ASSESSMENT QUESTIONS

44. The carotid pulse should be palpated

    A. with the client in a Fowler's position.
    B. when the client has been drinking.
    C. without using much pressure.
    D. as the client is sleeping.

45. The pulse volume

    A. is also referred to as pulse strength or amplitude.
    B. ranges from rhythmic to arrhythmic.
    C. may indicate heart rate.
    D. is measured with a pulsometer.

46. In an apical-radial pulse

    A. the pulses are usually taken by two nurses.
    B. both pulses are assessed by palpation.
    C. the apical and radial pulses are different in healthy adults.
    D. the radial pulse is usually greater than the apical.

47. The blood pressure can be taken in

    A. the arm over the brachial artery.
    B. the leg over the dorsalis pedis artery.
    C. the thigh over the popliteal artery.
    D. all of the above.

48. In a blood pressure reading

    A. there are three distinct sounds referred to as Korotkoff's sounds.
    B. the systolic pressure is the point where the second sound is heard.
    C. an auscultatory gap may be observed in hypertensive persons.
    D. the diastolic pressure is read first.

ADDITIONAL LEARNING ACTIVITIES

1. Practice assessing temperature, pulse, respirations and blood pressure as often as possible before carrying out the procedures in the clinical area.

2. Assess vital signs on an infant, child, adult and elderly person. What is the difference in assessing vital signs in each age group?

19 ASSESSING HEALTH STATUS

Assessment is the first and perhaps the most important aspect of the nursing process. When the assessment has been performed correctly, the nurse has the information necessary to identify a client's needs, develop appropriate nursing diagnoses and provide quality nursing care. This chapter focuses on the knowledge and skills the nurse needs to implement the first phase of the nursing process. After completing this chapter the student will be able to:

- Define terms associated with health assessment.
- Identify purposes of physical health assessment.
- Compare and contrast the four modes of physical assessment.
- Explain the significance of selected physical findings.
- Identify expected outcomes of health assessment.
- Identify the various steps in selected assessment procedures.
- Describe suggested sequencing to conduct a physical health assessment in an orderly fashion.

NURSING HEALTH HISTORY

What is included in the biographical data?

1. ______________________________

____

In a nursing health history, the chief complaint is:

2. ______________________________

The history of the present illness is a four-part elaboration of the chief complaint. It includes:

3. ______________________________

4. ______________________________

5. ______________________________

6. ______________________________

Why is the client's past history and family history of illness included in the nursing health history?

7. ______________________________

____

List the ten components of the nursing health history on the pages below.

NURSING HEALTH HISTORY

8.

9.

10.

11.

12.

13.

14.

15.

16.

17.

List the data that should be included in the past history.

18. ______________ 21. ______________

19. ______________ 22. ______________

20. ______________ 23. ______________

PHYSICAL HEALTH EXAMINATION

Describe the four methods the nurse uses to conduct the physical examination.

24. INSPECTION	
25. AUSCULTATION	
26. PERCUSSION	
27. PALPATION	

GENERAL SURVEY

Describe what the nurse is looking for when assessing the client's appearance and behavior in the following areas:

OBSERVATION	DESCRIPTION
28. POSTURE AND GAIT	
29. DRESS	
30. SIGNS OF DISTRESS	
31. ATTITUDE	
32. SPEECH	

INTEGUMENT

Match the terms in column I with the appropriate descriptions in column II.

33. ________ Hyperhidrosis
34. ________ Bromihidrosis
35. ________ Pallor
36. ________ Cyanosis
37. ________ Jaundice
38. ________ Vitiligo
39. ________ Albinism
40. ________ Ecchymosis

A. Whitish-gray-tinged skin
B. Patches of hypopigmented skin
C. Bruised skin
D. Excessive perspiration
E. Foul-smelling perspiration
F. Complete or partial lack of melanin
G. Blue-tinged skin
H. Yellow-tinged skin

HEAD

Label the bones of the head in the diagram on the left and the auscultation sites for bruits in the diagram on the right.

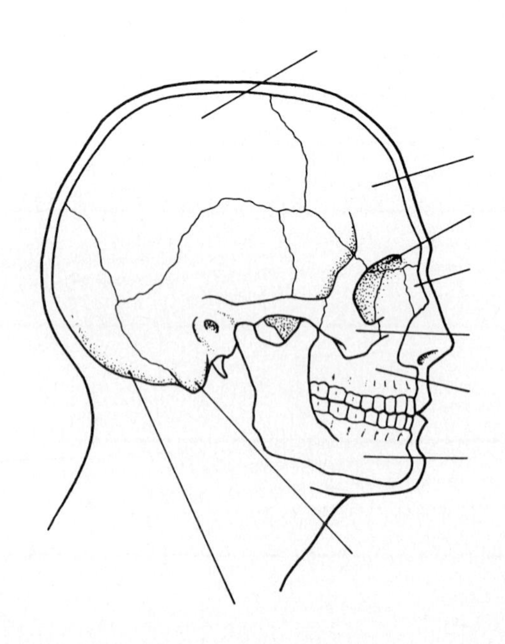

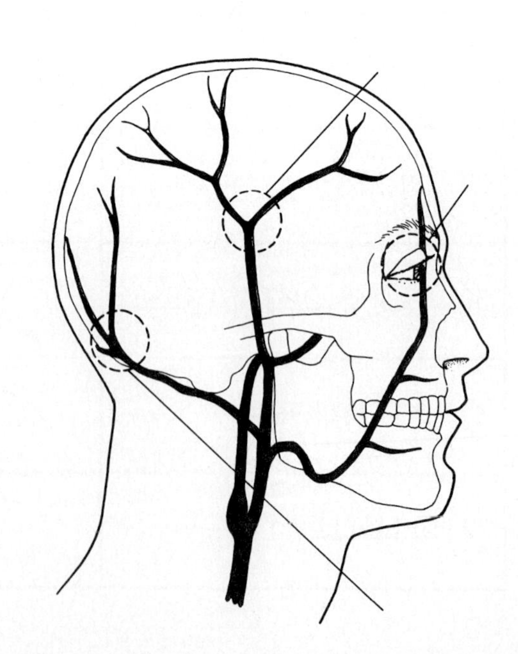

Label the structures of the neck in the diagram on the left. Label the lymph nodes of the neck in the diagram on the right.

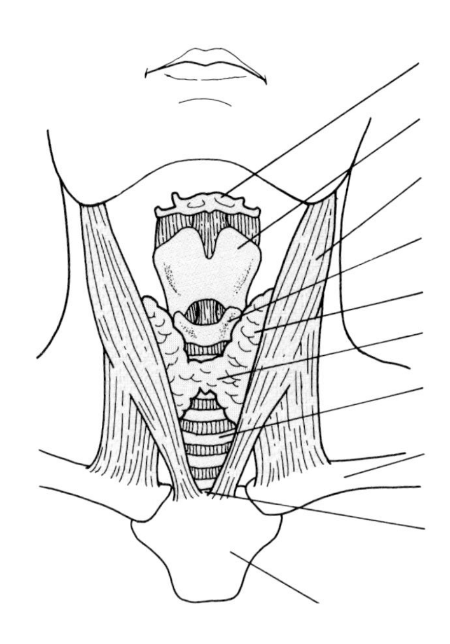

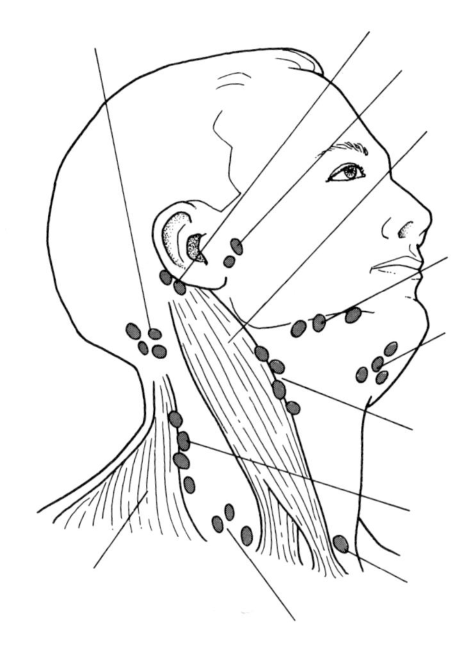

41. Another name for nearsightedness is ______________________________.

42. Another name for farsightedness is ______________________________.

43. Astigmatism can be defined as ______________________________.

44. Conjunctivitis is ______________________________.

45. Cataracts are ______________________________.

46. A lighted instrument used to inspect the tympanic membrane is an __________.

47. Bone conduction is assessed with the ______________________________ _ test.

48. The hard and soft palates are examined for ______________________________.

49. The lips are examined for ______________________________.

50. The ______________________________ is used to examine the nasal chambers.

51. Mixed hearing loss consists of _________________and _______________ loss.

Label the sinuses of the head in the diagrams below.

What kinds of assessment questions should the nurse ask the client when assessing the head and neck?

52. ____

____

____

What kinds of assessment questions should the nurse ask the client when assessing the eyes, ears, nose and mouth?

53. ____

____

THORAX AND LUNGS

In the diagrams below identify the following chest landmarks: the midsternal, left midclavicular, left and right anterior axillary, right midaxillary, and right posterior axillary lines.

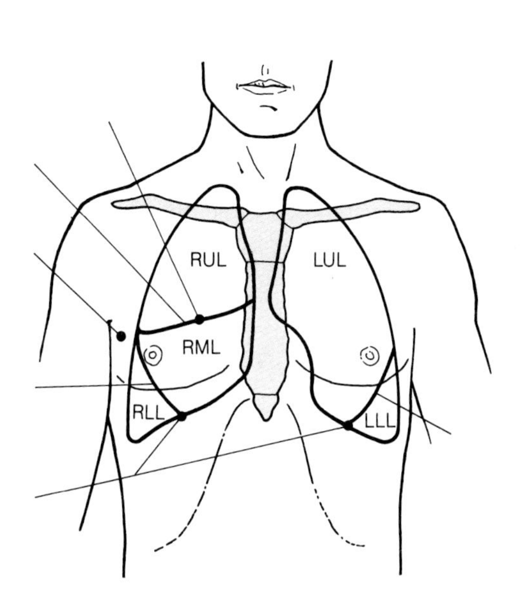

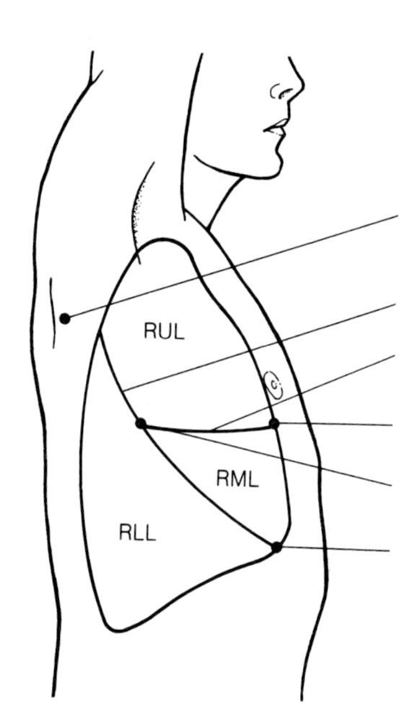

What kind of assessment questions should the nurse ask the client about the respiratory system?

54. ____

____

CARDIOVASCULAR SYSTEM

What kind of assessment questions should the nurse ask the client about the cardiovascular system?

55. ____

____

Label the areas over the precordium where aortic, pulmonic, tricuspid, and apical heart sounds are auscultated.

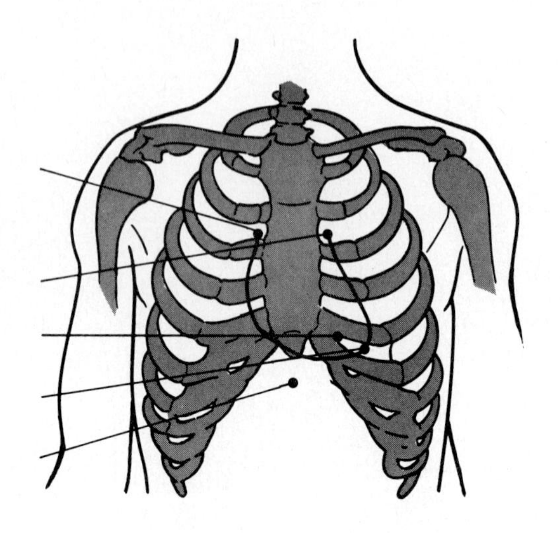

ABDOMEN

Identify the organs in the four abdominal quadrants in the table below.

UPPER RIGHT QUADRANT	UPPER LEFT QUADRANT
56.	57.
LOWER RIGHT QUADRANT	LOWER LEFT QUADRANT
58.	59.

What kinds of assessment questions should the nurse ask the client when assessing the abdomen?

60. ____

____

____

____

MUSCULOSKELETAL AND NEUROLOGICAL SYSTEM

What kinds of assessment questions should the nurse ask when assessing the musculoskeletal system?

61. ____

____

____

____

What kinds of assessment questions should the nurse ask when assessing the neurological system?

62. ____

____

____

____

FEMALE AND MALE GENITALS

Label the lymph node chains that are present in the female groin in the diagram below.

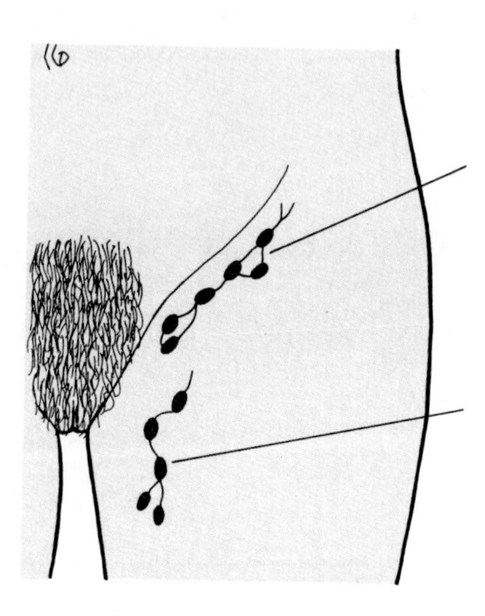

What kinds of assessment questions should the nurse ask the client when assessing the female genitalia and reproductive tract?

63. ______________________________

____

____

What kinds of assessment question should the nurse ask the client when assessing the male genitalia and reproductive tract?

64. ______________________________

____

____

SELF ASSESSMENT QUESTIONS

65. The best way to document a client's chief complaint is to

    A. utilize nursing terminology.
    B. write only objective data.
    C. use the client's own words.
    D. record it on the flowsheet.

66. When assessing the abdomen the nurse should

    A. auscultate, palpate, percuss and inspect.
    B. auscultate, percuss, palpate and inspect.
    C. inspect, auscultate, percuss and palpate.
    D. inspect, auscultate, palpate, and percuss.

67. When the nurse observes the client's affect, he or she is assessing the client's

    A. neurological status.
    B. musculoskeletal status.
    C. respiratory status.
    D. cardiovascular status.

68. Vesicular breath sounds are heard

    A. over the trachea.
    B. near the sternum.
    C. over the lower lung bases.
    D. over the heart.

69. The elderly client's abdomen is easier to examine than a younger client because

    A. the pain threshold of the client is decreased.
    B. the abdominal wall is slacker and thinner.
    C. muscle tone is increased.
    D. adipose tissue is decreased.

ADDITIONAL LEARNING ACTIVITIES

1. Use the nursing health history form in your clinical agency as a guide for interviewing a client.

2. Practice assessment techniques on a classmate in the clinical laboratory.

3. Visit a school health nurse. Observe how she or he assesses a child in this setting. What assessment tests are used?

4. Visit a nursing home. Select an elderly client and assess his or her emotional and intellectual status.

20 PREVENTING THE TRANSFER OF MICROORGANISMS

With the arrival of the HIV virus in the population and the increasing incidence of virulent microorganisms occuring within the general population in the last decade, nurses have become more knowledgeable than ever before about the strategies that must be used to protect the client from infection. This chapter focuses on assessment of risk factors that make a client vulnerable to infection as well as the nursing diagnoses, planning, and nursing strategies that are used in caring for and protecting the client. After completing this chapter the student will be able to:

- Describe the importance of biologic safety.
- Identify anatomic and physiologic barriers that defend the body against microorganisms.
- Describe the difference between nonspecific and specific defenses of the body.
- Differentiate active from passive immunity.
- Identify six links in the chain of infection.
- Identify measures that break each link in the chain of infection.
- Identify factors influencing a microorganism's capability to produce an infectious process.
- Identify people at risk of acquiring an infection.
- Describe four stages of an infectious process.
- Identify causal factors of nosocomial infections.
- Identify signs of localilzed and systemic infections.
- Identify relevant nursing diagnoses and contributing factors for clients at risk for infection and clients who have an infection.
- Develop outcome criteria to evaluate a client's response to nursing interventions and achievement of goals.
- Explain the concepts of medical and surgical asepsis.
- Identify interventions to prevent infections.
- Identify interventions to protect body defenses.
- Discuss methods for evaluating the effectiveness of protective measures.

IMPORTANCE OF BIOLOGIC SAFETY

Define the following terms associated with biologic safety.

1. Bacteriocins ______________________________

____

2. Flora ____

____

3. Virulence ____

____

4. Etiology ____

____

True and False

5. ______ A pathogen is a microorganism that produces disease.

6. ______ Inflammation is an adaptive mechanism that fights infection.

7. ______ In active immunity, the host produces its own antibodies in response to natural or artificial antigens.

8. ______ In passive immunity, the host receives natural or artificial antibodies produced by another source.

9. ______ Resident flora are microorganisms that are always present in and outside the body.

10. ______ Another name for immune defenses in nonspecific defenses.

11. ______ Lactoferrin is an iron-releasing protein that inhibits the growth of invading microorganisms.

12. ______ Colonization is the process by which strains of bacteria are removed from the body.

13. ______ Bacteria tend to reside in the moist dark areas of the body.

14. ______ Cilia are small hairlike projections that filter bacteria in the lungs.

15. ______ Hyperemia causes heat and redness at the site of an inflammatory process.

16. ______ Chemotaxis is the process whereby leukocytes are attracted to injured cells.

17. ______ Foreign proteins in the body are referred to as antigens.

CHAIN OF INFECTION

18. Identify the six links in the chain of infection by writing each one in a link in the chain below.

List two nursing interventions for each link of the chain of infection.

Etiological agent:

19. ______________________________

20. ______________________________

Source:

21. ______________________________

22. ______________________________

Portal of exit:

23. ______________________________

24. ______________________________

Mode of transmission:

25. ______________________________

26. ______________________________

Portal of entry:

27. ____________________

28. ____________________

Susceptible host:

29. ____________________

30. ____________________

FACTORS AFFECTING RISK OF INFECTION

List seven factors that affect host susceptibility.

31. ____________________
32. ____________________
33. ____________________
34. ____________________
35. ____________________
36. ____________________
37. ____________________

STAGES OF AN INFECTIOUS PROCESS

Describe the four stages of an infectious process in the boxes below.

NOSOCOMIAL INFECTIONS

What is a nosocomial infection?

42. ____

____

Why have health-care providers become concerned about nosocomial infections in recent years?

43. ____

____

ASSESSING

Identify the signs of localized infection and systemic infection in the table below.

LOCALIZED INFECTION	SYSTEMIC INFECTION
44.	49.
45.	50.
46.	51.
47.	52.
48.	53.

Potential for infection is the nursing diagnosis given to clients at risk for infection. Cite one risk factor NANDA uses as a defining characteristic of this diagnosis.

54. ______________________________

PLANNING CARE FOR SUSCEPTIBLE CLIENTS

Cite two outcome criteria that would be appropriate for a client with the nursing diagnosis potential for infection.

55. ______________________________

56. ______________________________

IMPLEMENTING NURSING STRATEGIES

List at least five interventions to prevent infections.

57. ______________________________

58. ______________________________

59. ______________________________

60. ______________________________

61. ______________________________

EVALUATING THE EFFECTIVENESS OF PROTECTIVE MEASURES

List three measures the nurse uses to evaluate whether or not the client has achieved the goals established to prevent or overcome infectious processes.

62. ______________________________

63 ______________________________

64. ______________________________

SELF ASSESSMENT QUESTIONS

65. Biological safety is important in nursing practice because

    A. nurses are directly involved in maintaining a safe environment.
    B. pathogenic bacteria are everywhere.
    C. nosocomial infections are becoming more virulent.
    D. all of the above are true.

66. Which of the following is not a source of nosocomial infections?

    A. Endogenous sources
    B. Exogenous sources
    C. Iatrogenic sources
    D. Progenic sources

67. Which of the following is not a true statement? During an infectious process

    A. the white blood cell count rises to 4,500 to 11,000/cu mL.
    B. the erythrocyte sedimentation rate is elevated.
    C. the neutrophil count is elevated if the infection is suppurative.
    D. the eosinophil count usually remains normal.

68. Objects used in a sterile field

    A. may contain nonpathogenic organisms.
    B. can be stored indefinitely.
    C. should never be left unattended after opening.
    D. do not always remain sterile.

69. After giving an injection, the needle should be

    A. recapped and discarded in a garbage container.
    B. left uncapped and discarded in a designated receptacle.
    C. washed and resterilized for future use.
    D. placed in a disinfecting solution until the next use.

ADDITIONAL LEARNING ACTIVITIES

1. Interview an infection control nurse epidemiologist from a local clinical facility. What are the functions this nurse performs? What statistical data related to infection control are collected?

2. Observe the health care providers on your clinical unit. Are infection control procedures such as hand washing, bagging of linens or isolation procedures being carried out?

21 PROVIDING A SAFE ENVIRONMENT

Since accidents are a major cause of death among individuals of all ages in society today, Chapter 21 focuses on the responsibility of the nurse in providing a safe environment for the client, whether it be at home or in a clinical setting. Special emphasis is placed on the major causes of accidental deaths that can occur in the hospital or at home: falls, fires, poisoning, suffocation, radiation, and electrical shock. Assessment for risk factors related to potential accidents is reviewed as well as the diagnosis, planning, implementation, and evaluation of nursing measures to prevent them. After reviewing this chapter the student will be able to:

- Identify clients at risk of physical injury.
- Identify common hazards in the home.
- Give examples of nursing diagnoses for clients at risk for accidental injury.
- List outcome criteria for evaluating selected strategies for preventing injury.
- Describe nursing responsibilities regarding fires.
- Identify common causes of scalds and burns in home and hospital settings.
- Identify precautions to prevent falls of hospitalized clients.
- Describe legal implications of restraining clients.
- State guidelines for selecting and applying restraints.
- Identify essential precautions to prevent poisoning.
- Identify measures to reduce electrical hazards.
- Describe measures to minimize noise.
- List precautions to prevent exposure to radiation.

ASSESSING CLIENTS AT RISK FOR INJURY

Marjory Evan is a 72-year-old widow who lives alone in the home she and her husband shared for fifty years. Although Marjory's home is in a high-crime area, she refuses to move because she says, "I've lived here for fifty years, raised a family and buried a husband. I'm not leaving now!" Marjory's children are concerned because she is hard of hearing, forgetful, and has had several accidents in the last six months. Last week one of her daughers found Marjory asleep in her recliner chair with a lit cigarette in her hand. In the last six months Majory has complained of stiffness in her hips and back and has fallen several times because she has trouble keeping her balance when going up and down the stairs. Marjory's daughters are concerned that she will be injured because of her many problems.

List three risk factors that make Marjory a potential victim of injury.

1. ______________________________

2. ______________________________

3. ______________________________

Marjory's home is old and in need of repair. List five factors that should be considered when appraising an elderly person's home for safety hazards.

4. ______________________________

5. ______________________________

6. ______________________________

7. ______________________________

8. ______________________________

DIAGNOSING AND PLANNING

Marjory's daughter enlists the help of a home-health-agency nurse to assess her mother's needs. If you were the home health nurse that visits Marjory, what nursing diagnoses would you develop for her? List at least two nursing diagnoses for Marjory in the table below, then develop one outcome criteria/objective for each nursing diagnosis.

NURSING DIAGNOSES	OUTCOME CRITERIA/OBJECTIVES
9.	11.
10.	12.

IMPLEMENTING STRATEGIES IN RESPONSE TO SPECIFIC HAZARDS

One afternoon, Marjory fell and injured her back while sweeping the basement steps. She was admitted to the orthopedic unit of the hosptial. Although there is a nonsmoking policy on the unit, Marjory continues to smoke. One afternoon Marjory's nurse finds her smoking in bed. The nurse tell Marjory that three elements have to be present to support a fire. They are:

13. ______________________________

14. ______________________________

15. ______________________________

The nurse goes on to tell Marjory that cigarette smoking is a common cause of burns in elderly clients. List two hazards that cause scalds and two hazards that cause burns in clients.

SCALD HAZARDS	BURN HAZARDS
16.	18.
17.	19.

Marjory continues to display weakness and lack of balance in the hospital. Describe five guidelines the nurse should consider to prevent a client like Marjory from falling.

20. ______________________________

21. ______________________________

22. ______________________________

23. ______________________________

24. ______________________________

Occasionally Marjory gets confused during the night and needs to be restrained to prevent her from falling and hurting herself again. What are the legal ramifications of placing Marjory in restraints?

25. ______________________________

____

List the guidelines the nurse should follow when selecting a restraint for Marjory.

26. ______________________________

27. ______________________________

28. ______________________________

29. ______________________________

30. ______________________________

Marjory is placed in a body restraint. What guidelines should the nurse observe while Marjory is in the restraint?

31. ______________________________

32. ______________________________

33. ______________________________

34. ______________________________

35. ______________________________

Describe three precautions to prevent poisoning and three precautions to prevent electrical hazards.

POISONING PRECAUTIONS	ELECTRICAL PRECAUTIONS
36.	38.
37.	39.

Describe two measures to minimize noise.

40. ______________________________

41. ______________________________

Describe two measures to prevent exposure to radiation.

42. ______________________________

43. ______________________________

SELF ASSESSMENT QUESTIONS

44. A hallucination is

    A. the perception of external stimuli when none exists.
    B. misinterpretation of external stimuli.
    C. the ability to perceive environmental stimuli.
    D. sensory input from the environment.

45. In the event of a fire, the nurse should

    A. close windows, use the fire extinguisher, and evacuate clients.
    B. call the switchboard, activate the fire alarm, and use the fire extinguisher.
    C. evacuate the clients, use the fire extinguisher, and activate the fire alarm.
    D. Evacuate the clients, activate the fire alarm, and call the switchboard.

46. To prevent poisoning, teach parents to

    A. lock cleaning agents in a cupboard where children cannot reach.
    B. place poison warning stickers designed for children on toxic containers.
    C. avoid storing toxic liquids or solids in food containers.
    D. do all of the above.

47. To prevent electical hazards, the nurse should teach the client to

    A. use noninsulated wiring that does not conduct electricity.
    B. disconnect appliances before cleaning, since water conducts electricity.
    C. use UL approved electrical appliances around sinks and bathtubs.
    D. use no more than four extension cords in an outlet.

48. In order to prevent falls in elderly clients, the nurse should

    A. encourage the client to use footwear with smooth soles.
    B. discourage the use of bath mats in tubs and showers.
    C. keep the bed in the high position.
    D. attach side rails to the client's bed.

ADDDITIONAL LEARNING ACTIVITIES

1. Gather statistics from the safety officer in the clinical agency regarding types and numbers of accidents that have occurred over the last year. How many accidents happened to clients? Employees?

2. Analyze the data you collected in number 1. How could each of the accidents have been prevented?

3. Appraise your own home for safety hazards. What do you need to do to remove these hazards from your home?

22 HYGIENE

The provision and maintenance of personal hygiene for the client is one of the most important tasks that nurses perform. Although hygienic procedures may seem very basic, the professional nurse knows that they give the nurse an opportunity to promote cleanliness and comfort, improve circulation, prevent complications of bedrest, establish a nurse/client relationship, and assess the client's health status. This chapter includes the knowledge and skills the nurse needs to provide personal hygienic measures for the client. After completing this chapter the student will be able to:

- Describe kinds of hygienic care a nurse provide to clients.
- Identify factors influencing personal hygiene.
- Identify normal and abnormal findings obtained during inspection and palpation of the skin, feet, nails, mouth, hair, eyes, ears, and nose.
- Describe variations in the appearance of the skin, nails, and mucous membranes of light-skinned and dark-skinned clients.
- Identify common problems of the skin, feet, nails, mouth, hair, eyes, ears, and nose and formulate related nursing diagnoses.
- Describe guidelines for planning and implementing nursing interventions for the skin, feet, nails, mouth, hair, eyes, ears, and nose.
- List outcome criteria to evaluate goal achievement.
- Identify the purposes of bathing.
- Describe various types of baths.
- Describe steps in perineal and genital care.
- Explain five techniques used in back rubs.
- Explain specific ways in which nurses help hospitalized clients with oral hygiene.
- Identify steps in inserting and removing contact lenses and artificial eyes.
- Describe steps in inserting and removing hearing aids.
- Identify safety and comfort measures underlying bedmaking procedures.

HYGIENIC CARE

List the four purposes of hygiene.

1. ____________________

2. ____________________

3. ____

4. ____

Describe the various types of hygiene care clients commonly receive.

5. Early morning care ____

____

6. Morning care ____

____

7. Afternoon care ____

____

8. Hour of sleep ____

____

SKIN

Describe the skin problem listed in the table below, then describe the nursing implications.

SKIN PROBLEM AND DESCRIPTION	NURSING IMPLICATIONS
9. Abrasion	
10. Acne	
11. Erythema	

List the conditions that place a client at risk for skin impairments or breakdown.

12. ____________________

13. ____________________

14. ____________________

15. ____________________

16. ____________________

17. ____________________

18. ____________________

Develop two nursing diagnoses for the client who is immobilized. Then develop at least one outcome criteria and one nursing intervention for each diagnosis.

NURSING DIAGNOSIS	CLIENT GOAL/ OUTCOME CRITERIA	NURSING INTERVENTION AND RATIONALE
19.		
20.		

List six guidelines for skin care that the nurse uses when caring for clients.

21. ____________________

____

22. ____________________

____

23. ____

____

24. ____

____

25. ____

____

26. ____

____

Draw the direction the hands move when massaging a client's back in the diagram below.

FEET

Draw a line from the foot problem in column I to its description in column II.

27. Callus
28. Corn
29. Plantar warts
30. Fissure
31. Athlete's foot
32. Ingrown toenail

A. Keratosis caused by friction and pressure from a shoe

B. Growths caused by the virus papovavirus hominis

C. The growing inward of the nail into the soft tissues around the nails

D. A thickened portion of epidermis consisting of a mass of keratotic material

E. Tinea pedis or ringworm of the foot

F. A deep groove that occurs between the toes

Label the joints of the large toe in the diagrams below, then identify the types of toe deviations depicted in the space provided.

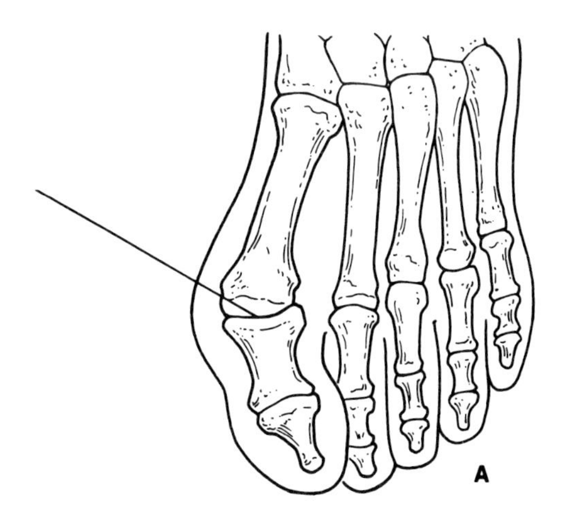

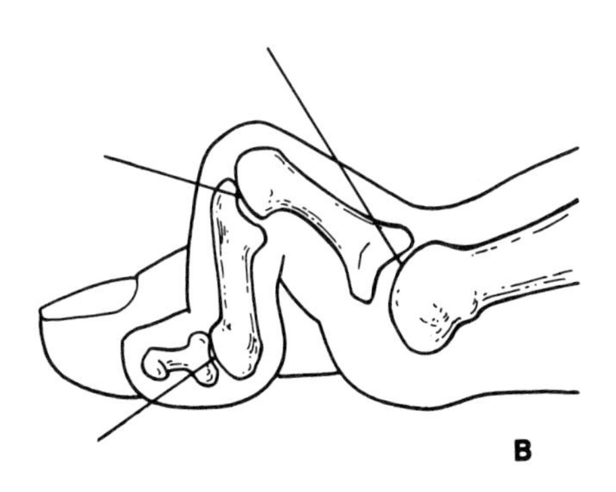

33. A is a diagram of____________________________________

34. B is a diagram of ____________________________________

What is the nurse looking for when assessing the feet? What plantar landmarks should be noted?

35. ____

____

____

____

Develop two nursing diagnoses for the client who has foot problems. Then develop at least one outcome criteria and one nursing intervention for each diagnoses.

NURSING DIAGNOSIS	CLIENT GOAL/ OUTCOME CRITERIA	NURSING INTERVENTION AND RATIONALE
36.		
37.		

NAILS

Describe the care a nurse should give to a client who is unable to provide his or her own nail care.

38. ____

____

HAIR

True and False

39. _______ In severe cases, dandruff affects the auditory canals and eyebrows.

40. _______ Rocky Mountain spotted fever and tularemia are two diseases transmitted by ticks.

41 _______ Pediculus capitus is a type of lice found in the hair, pubic area, and clothing.

42. _______ Hirsuitism is the growth of excessive body hair.

43. _______ Although the shafts of spiraled or curly hair look strong, they have less strenth than straight hair shafts and are easily broken.

44. _______ A client should never shampoo the head in the shower.

45. _______ An electric shaver should always be used when shaving a client who is ill.

EYES, EARS, NOSE, AND MOUTH

MOUTH PROBLEM AND DESCRIPTION	NURSING IMPLICATIONS
46. Halitosis	
47. Chilosis	
48. Periodontal disease	

Describe how to insert contact lenses in the eyes of a client who is unable to do so for himself or herself.

49. ____

____

____

Describe the steps for administering mouth care for an unconscious client.

50. ____

____

____

Develop one nursing diagnosis for each of the following client problems: mouth eyes, ears, and nose. Then develop at least one outcome criteria and one nursing intervention for each diagnosis.

NURSING DIAGNOSIS	CLIENT GOAL/ OUTCOME CRITERIA	NURSING INTERVENTION AND RATIONALE
51. Mouth problems:		
52. Eye problems:		
53. Ear problems:		
54. Nose problems:		

SELF ASSESSMENT QUESTIONS

55. When washing the client's extremities, the nurse uses long, firm strokes, moving from distal to proximal areas to

    A. provide range of motion.
    B. assess the skin.
    C. improve circulation.
    D. prevent infection.

56. The eyes should be washed from the inner canthus to the outer canthus to

    A. avoid the nose.
    B. decrease friction.
    C. promote circulation.
    D. prevent infection.

57. Whenever a client's dentures are removed, they should be

    A. sent home with the family.
    B. placed in a denture cup.
    C. marked and put in a drawer.
    D. wrapped in tissue.

58. The unconscious client should be placed in a side-lying position for mouth care to

    A. prevent aspiration.
    B. promote rest.
    C. increase awareness.
    D. encourage coughing.

59. The bottom sheets are always kept as wrinkle free as possible to

    A. provide neatness.
    B. decrease backaches.
    C. prevent pressure sores.
    D. increase sacral circulation.

ADDITIONAL LEARNING ACTIVITIES

1. Develop and implement a teaching plan for oral hygiene for preschool through grade school students at a local grade school.

2. Plan a health fair for a local senior citizen facility with your classmates. Develop exhibits focusing on the hygienic health needs of older adults.

3. Develop and implement a plan of care for a client who needs complete hygienic care.

23 HEALTH PROMOTION

In the not-so-far distant past, nurses were primarily interested in the care of the sick, infirm, or the dying within the confines of the hospital or nursing home. The primary focus of nursing care at that time was disease *cure* rather than disease *prevention*. with the increased knowledge and technology of the past few decades, however, nurses have bcome increasingly aware of their role of assisting the client attain his or her highest level of health in primary, secondary, and tertiary health-care settings. This chapter assists the student in incorporating health-promotion skills into the nursing process. After completing this chapter the student will be able to:

- Explain the essential facts about health promotion.
- Compare Pender's concept of health promotion with that of Leavell and Clark.
- List the various types of health-promotion programs.
- Describe the common sites for health-promotion activities.
- Discuss the nurse's role in health-promotion.
- Discuss the steps of the health-promotion process.
- Identify the essential components of health-promotion assessment.
- Compare wellness nursing diagnoses with NANDA nursing diagnoses.
- Explain the steps involved in health-promotion planning.
- Discuss nursing strategies for providing and facilitating support.
- List the essential guidelines for enhancing behavior change.
- Describe the evaluation phase of the health-promotion process.

CONCEPT AND SCOPE OF HEALTH PROMOTION

Define each of the following theorist's definiton of health promotion.

THEORIST	HEALTH PROMOTION DEFINITION
Pender(1987)	1.
Leavell and Clark(1965)	2.
Richmond(1979	3.
Maglacas(1988)	4.

List and briefly describe the various types of health-promotion programs and the sites where they are being implemented.

5. ______________________________

6. ______________________________

7. ______________________________

8. ______________________________

9. ______________________________

List at least five roles the nurse plays in health promotion.

10. ______________________________

11. ______________________________

12. ______________________________

13. ______________________________

14. ______________________________

List the steps of the nursing process in health promotion in the spaces below.

15. ______________________________

16. ______________________________

17. ______________________________

18. ______________________________

19. ______________________________

ASSESSING

Describe six components of health-promotion assessment in the spaces below.

20. ______________________________

21.

22.

23.

24.

25.

DIAGNOSING

26. Describe the difference between nursing diagnoses related to wellness and NANDA nursing diagnoses related to health promotion in the appropriate boxes below.

NURSING DIAGNOSES RELATED TO WELLNESS	NANDA DIAGNOSES RELATED TO HEALTH PROMOTION

PLANNING AND IMPLEMENTING

List the seven steps that Pender(1987) outlines in the process of health planning.

27.

28.

29.

30.

31.

32.

33.

Describe four stategies the nurse can use to support the client.

34. ______________________________

35. ______________________________

36. ______________________________

37. ______________________________

How does the nurse use Lewin's stages of change to enhance behavior change.

38. ______________________________

____

____

____

EVALUATING

Discuss what happens during the evaluation phase.

39. ____

____

SELF ASSESSMENT QUESTIONS

40. Health promotion is (are)

    A. activities directed towards increasing the level of well-being and actualizing the potential of the client.
    B. maintaining or improving the general level of health of the client.
    C. individual and community activities to promote healthful life-styles.
    D. all of the above.
    E. none of the above.

41. In order to enhance behavior aimed at improving or maintaining health in the client, the nurse should

    A. ask the client to explain his or her lack of motivation.
    B. recognize that the client may be resistent to change,.
    C. dismiss the client's cultural background.
    D. insist that the client change unhealthy behavior.

42. The best definition of a risk factor is

    A. a life-style behavior that increases a person's chance of acquiring a specific disease.
    B. a phenomenon that occurs when a person reduces the probability of acquiring a disease.
    C. a condition that occurs when a person places him or herself in an environmental hazard.

43. The diagnosis which is *most* geared toward wellness is

    A. ineffective coping related to a terminal illness.
    B. alteration in breathing patterns related to environmental pollution.
    C. alteration in nutrition less than body requirements.
    D. Potential for physical fitness related to motivation.

44. A nurse will be more effective in assisting the client to change unhealthy habits if the nurse models healthy life-style behaviors and attitudes in his or her life. This statement is

    A. true.
    B. false.

ADDITIONAL LEARNING ACTIVITIES

1. Obtain a copy of a health risk appraisal form. Use the form to assess your own risk factors. Make a list of the life-style changes you will need to implement after completing the form.

2. Complete the health risk appraisal form on an older and younger member of your family. Compare the similarities and differences.

3. Attend a session of a health-promotion program in your community. What risk factors does this program address? What stategies do the leaders of this program use to motivate their clients?

24 CONCEPTS OF GROWTH AND DEVELOPMENT

The concepts of growth and development and maturation are independent, interrelated processes that are an integral part of holistic nursing care. In nursing, these theories can be helpful in guiding assessment, explaining behavior, and providing a direction for nursing interventions. This chapter presents an overview of the major growth, development, and maturation theories that nurses commonly use for client care. After completing this chapter the student will be able to:

- Describe essential facts related to growth and development.
- Differentiate growth, development, and maturation.
- Describe the stages of growth and development.
- List factors that influence growth and development.
- Explain the principles of growth and development.
- Differentiate Gesell's and Havighurst's theories of development.
- Compare Freud's theory of psychosexual development with Sullivan's theory of interpersonal development.
- Differentiate Skinner's classical and operant conditioning.
- Describe essential aspects of Bandura's social learning theory.
- Compare Peck's and Gould's stages of adult development.
- Explain Piaget's theory of cognitive development.
- Compare Kolberg's, Peter's, and Gilligan's theories of moral development.
- Compare Fowler's and Westerhoff's stages of spiritual development.

GROWTH, DEVELOPMENT, AND MATURATION

Describe at least five of the principles related to growth and development.

1. ______________________________

2. ______________________________

3. ______________________________

4. ______________________________

5. ______________________________

Describe the processes of growth, development and maturation in the boxes below.

6. Growth
7. Development
8. Maturation

Describe three genetic factors that influence growth and development.

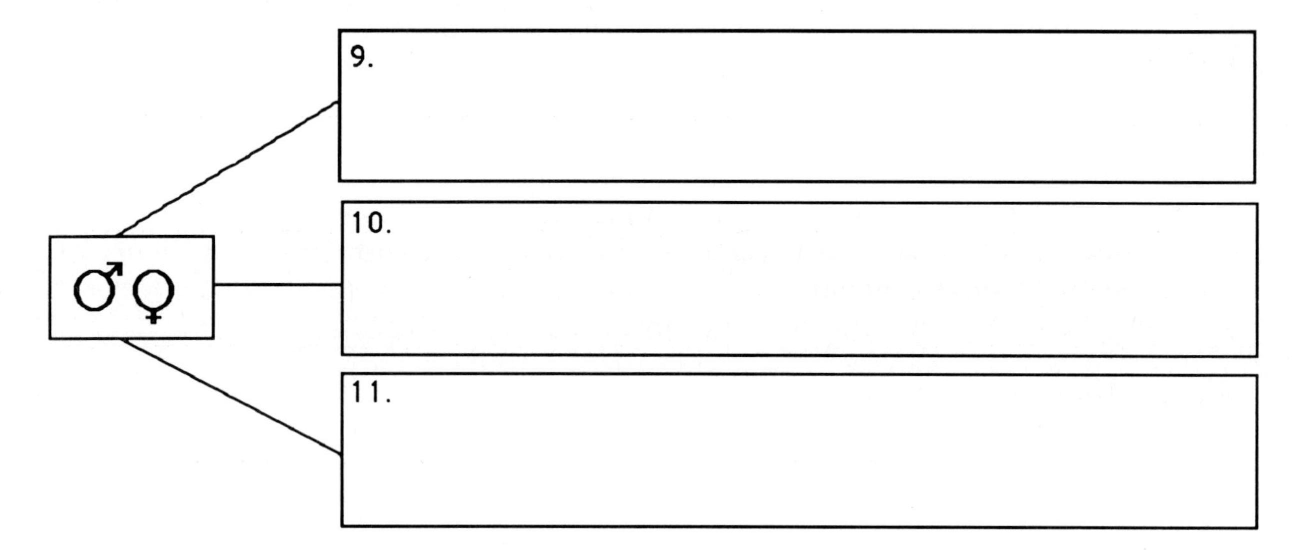

Describe three environmental factors that influence growth and development.

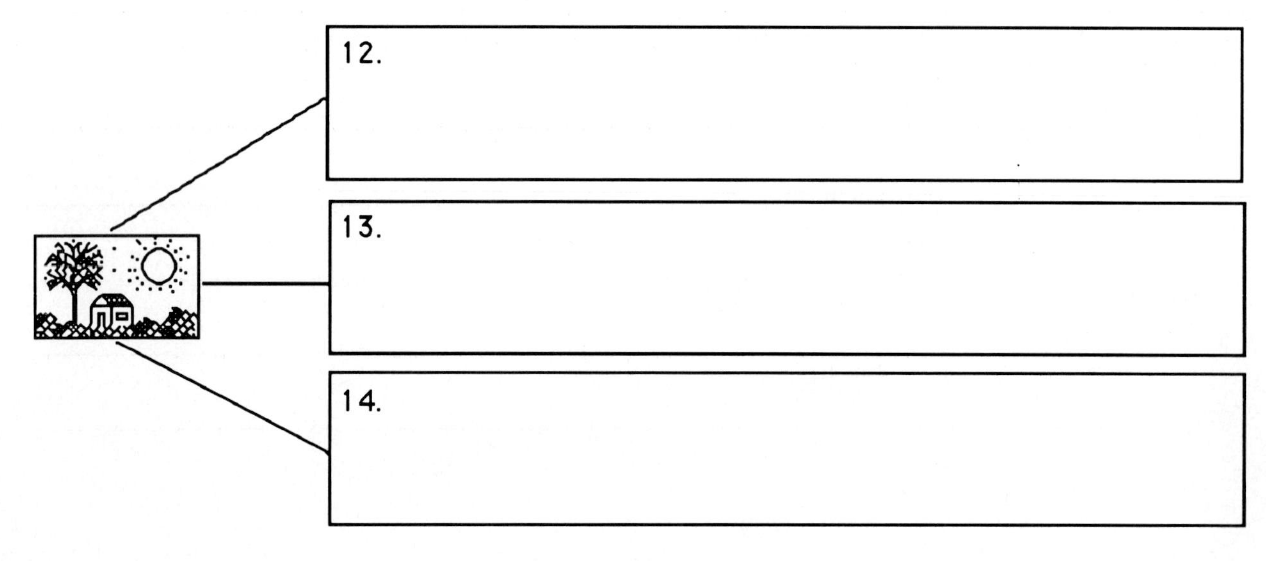

15. Fill in the blank areas of the growth and development table that follows.

STAGE	AGE	SIGNIFICANT CHARACTERISTICS	NURSING IMPLICATIONS
Neonatal			
Infancy			
Toddlerhood			
Preschool			
School Age			
Adolescence			
Young Adulthood			
Middle Adulthood			
Young Old Adulthood			
Middle Old Adulthood			
Old Old Adulthood			

MATURATION AND PSYCHOSOCIAL THEORIES

Briefly explain the developmental theories for each pair of theorists in the spaces provided. Then think about how each differs and how they are similar.

16. Maturation theories:

Gesell	Havighurst

17. Psychosocial theories

Freud	Sullivan

Identify the terms in Column I by placing the appropriate letter from Column II in the space provided.

18. _____ Response
19. _____ Classical conditioning
20. _____ Operant conditioning
21. _____ Reinforcement
22. _____ Behavior modification
23. _____ Extinction

A. Positive, desirable actions are praised and rewarded. Undesirable actions are ignored.
B. Conditioned responses are established by the association of a new stimulus that is known to cause and unconditioned response.
C. Consequences of an action.
D. An act that can be traced to the effects of a stimulus.
E. The frequency of a response is increased or decreased depending on when, how, and to what extent it is reinforced.
F. A conditioned behavior is "unlearned" because the reinforcement has been removed.

Briefly describe the essential aspects of Bandura's social learning theory.

24. ____

____

25. Briefly explain Gould and Peck's theories of adult development, then take a minute to think about their theories and how they relate to each other.

Gould	Peck

COGNITIVE THEORY

Briefly describe Piaget's cognitive theory.

26. ____

____

MORAL THEORIES

27. Briefly explain Kolberg's, Peter's and Gilligan's theories of moral development, then take a few minutes to consider how their theories compare.

Kolberg	Peter

Gilligan

SPIRITUAL THEORY

28. Briefly explain Fowler's and Westerhoff's stages of spiritual development, then take a few minutes to consider how their theories compare.

Fowler	Westerhoff

SELF ASSESSMENT QUESTIONS

29. Toilet training is a major developmental task of the

    A. infant.
    B. toddler.
    C. preschooler.
    D. none of the above.

30. According to Erikson, a school-age child is struggling with

    A. trust vs. mistrust.
    B. autonomy vs. shame and doubt.
    C. industry vs. inferiority.
    D. self-identity vs. role confusion.

31. According to Sullivan a person begins to integrate self-esteem in

    A. infancy.
    B. childhood.
    C. preadolescence.
    D. late adolescence.

32. Maturation is (are)

    A. physical changes in size.
    B. genetically influenced physical changes.
    C. the increase in the complexity of function and skill progression.
    D. the rate of development

33. Developmental theories are useful in nursing because they

    A. guide assessment.
    B. explain client behaviors.
    C. influence nursing interventions.
    D. include all of the above.

ADDITIONAL LEARNING ACTIVITIES

1. Visit a family with several children in different stages of growth and development. Compare the growth and development of the children using Erikson's eight stages of development.

2. Interview an adolescent and an elderly person. Compare their interpersonal, physical, spiritual, and moral development.

25 INFANCY THROUGH LATE CHILDHOOD

Chapter 25 focuses on the growth and development of the infant, toddler, preschooler, and school-aged child. The physical, psychosocial, cognitive, moral, and spiritual aspects that occur in each stage are discussed as well as the developmental tasks that are characteristic of each stage. After completing this chapter the student will be able to:

- Identify characteristic tasks at different stages of development during infancy and childhood.
- Describe usual physical development throughout infancy and childhood.
- Trace psychosocial development according to Erikson through infancy and childhood.
- Explain changes in cognitive development according to Piaget throughout infancy and childhood.
- Describe moral development according to Kohlberg throughout childhood.
- Describe spiritual development according to Fowler throughout childhood.
- Identify assessment activities and expected characteristics from birth through late childhood.
- Identify nursing diagnoses for health promotion from birth through late childhood.
- List examples of health promotion goals from birth through late childhood.
- Identify essential health promotion and protection activities to meet the needs of infants, toddlers, preschoolers, and school-age children.

ASSESSMENT ACTIVITIES

List three nursing assessment activities for each stage of growth and development from infancy to late childhood.

Infancy:

1. ______________________________

2. ______________________________

3. ______________________________

Toddlers:

4. ______________________________

5. ______________________________

6. ______________________________

Preschoolers:

7. ______________________________

8. ______________________________

9. ______________________________

School-age:

10. ______________________________

11. ______________________________

12. ______________________________

DEVELOPMENTAL ASPECTS OF INFANCY AND CHILDHOOD

Briefly describe the physiologic, psychosocial, and cognitive aspects of each stage of growth and development in the table below.

STAGE	PHYSIOLOGIC	PSYCHOSOCIAL	COGNITIVE
13. Infancy			
14. Toddler			
15. Preschool			
16. School-age			

Describe the developmental tasks and the spiritual and moral aspects of each stage below.

STAGE	TASK	SPIRITUAL	MORAL
17. Infancy			
18. Toddler			
19. Preschool			
20. School-age			

HEALTH PROMOTION AND PROTECTION

Cite an example of a health promotion nursing diagnosis and a related goal for each stage of growth and development.

INFANCY

21. Nursing diagnosis ______________________________

22. Goal______________________________

TODDLER

23. Nursing diagnosis ______________________________

24. Goal______________________________

PRESCHOOL

25. Nursing diagnosis______________________________

26. Goal______________________________

SCHOOL - AGE

27. Nursing diagnosis____________________

28. Goal____________________

Indicate the ages when each of the following immunizations should be given to a child from birth through the 16th year.

29. Diphtheria-pertussis-tetanus ____________________

30. Poliomyelitis ____________________

31. Measles-mumps-rubella____________________

SELF ASSESSMENT QUESTIONS

32. By six months of age an infant

    A. walks with help.
    B. sits with support.
    C. uses a spoon.
    D. waves bye-bye.

33. 20-20 vision is well established by

    A. 1 year.
    B. 5 years.
    C. 10 years.
    D. 15 years.

34. The central task of the school-age child according to Erickson is

    A. trust vs. mistrust.
    B. industry vs. inferiority.
    C. self-identity vs. role confusion.
    D. autonomy vs. shame.

35. When a toddler has a tantrum the mother should

    A. ask the child why he or she is angry.
    B. give the child what she or he wants.
    C. spank the child.
    D. make sure the child is safe then leave.

36. A five year old should be able to do all of the following *except*

    A. throw and catch a small ball.
    B. jump rope and skip.
    C. climb playground equipment.
    D. ride a bicycle with training wheels.

ADDITIONAL LEARNING ACTIVITIES

1. Observe a newborn infant at home. Is there any communication between the mother and infant?

2. Visit a local preschool to observe the children's behavior during various activities throughout the day.

3. Interview a school-aged child. Determine whether he or she is mastering the developmental tasks appropriate for this age group.

26 ADOLESCENCE THROUGH MIDDLE ADULTHOOD

Chapter 26 focuses on the growth and development of the adolescent, young adult, and middle aged adult. The physical, psychosocial, cognitive, moral, and spiritual aspects that occur in each stage are discussed as well as the developmental tasks that are characteristic of each stage. After completing this chapter the student willbe able to:

- Explain the essential changes in physical development from adolescence through middle adulthood.
- Explain psychosocial development of adolescents, young adults, and middle-aged adults according to Erikson.
- Explain the essential changes in cognitive development from adolescence through middle adulthood as postulated by Piaget.
- Describe moral development of adolescents, young adults, and middle-aged adults according to Kohlberg.
- Discuss spiritual development of adolescents, young adults, and middle-aged adults according to Fowler.
- Identify common health hazards and concerns of adolescents, young adults, and middle-aged adults.
- Discuss nursing implications related to common health concerns identified.

ASSESSMENT ACTIVITIES

List three nursing assessment activities for each stage of growth and development from adolescence to middle-aged adulthood.

Adolescence:

1. ______________________________

2. ______________________________

3. ______________________________

Young adults:

4. ______________________________

5. ______________________________

6. ______________________________

Middle-age:

7. ______________________   9. ______________________

8. ______________________

DEVELOPMENTAL ASPECTS

Briefly describe the physiologic, psychosocial, and cognitive aspects of each stage.

STAGE	PHYSIOLOGIC	PSYCHOSOCIAL	COGNITIVE
10. Adoles-cent-			
11. Young adult			
12. Middle-aged adult			

Briefly describe the developmental tasks and the spiritual and moral aspects of each stage of growth and development.

STAGE	TASK	SPIRITUAL	MORAL
13. Adoles-cent-			
14. Young adult			
15. Middle-aged adult			

HEALTH PROMOTION AND PROTECTION

Cite an example of a health promotion nursing diagnosis and a related goal for each stage of growth and development.

ADOLESCENCE

16. Nursing Diagnosis ____________________

17. Goal____________________

YOUNG ADULT

18. Nursing diagnosis ____________________

19. Goal____________________

MIDDLE - AGED ADULT

20. Nursing diagnosis____________________

21. Goal____________________

Identify at least two potential health hazards associated with each stage from adolescence to middle-age.

ADOLESCENCE

22. ____________________

23. ____________________

YOUNG-ADULT

24. ____________________

25. ____________________

MIDDLE-AGED ADULT

26. ____________________

27. ____________________

NURSING IMPLICATIONS

Describe at least two nursing implications when caring for a client in adolescence, young adulthood, and middle-aged adulthood.

ADOLESCENCE

28. ____

29. ____

YOUNG-ADULT

30. ____

31. ____

MIDDLE - AGED ADULT

32. ____

33. ____

SELF ASSESSMENT QUESTIONS

34. An adolescent may fear a surgical procedure because it

    A. may interfere with homework.
    B. could alter his or her self-concept.
    C. means a general anesthetic is used.
    D. will necessitate intravenous therapy.

35. In late adolescence the teen-ager

    A. is concerned about the sudden physical changes that are being experienced.
    B. plans for the future and economic independence.
    C. is at the conventional level of moral development.
    D. experiences all of the above.

36. Toxic shock syndrome (TSS)

    A. is related to premenstrual syndrome (PMS).
    B. may be associated with tampon use.
    C. occurs between the ages of 12 and 15 years.
    D. is related to dysmenorrhea.

37. Testicular self-examination

    A. is recommended in males over 30 years old.
    B. should only be done in males with a history of cancer in their family.
    C. should be conducted monthly.
    D. should be done before taking a bath or shower.

38. In general the middle-aged adult

    A. has decreased capacity to learn.
    B. has less time for leisure activities.
    C. begins to feel the affects of aging.
    D. begins to experience memory loss.

ADDITIONAL LEARNING ACTIVITIES

1. Interview an adolescent, young adult, and a middle-aged adult. Compare their developmental needs and problems. How are they alike? How do they differ?

2. Compare your own growth and development to the standards outlined in this chapter for physical, psychosocial, cognitive, spiritual, and moral development.

3. Prepare a teaching plan for breast or testicular self-examination. Teach the procedure to a family member or friend.

4. Begin to carry out breast or testicular self-examination (whichever is appropriate) for yourself if you have not done so in the past.

27 LATE ADULTHOOD

Chapter 27 focuses on the growth and development of the client in late adulthood. The physical, psychosocial, cognitive, moral and spiritual aspects that occur during this stage is discussed as well as characteristic developmental tasks. After completing this chapter the student will:

- Describe the physical changes that occur from middle adulthood through old age.
- Explain essential aspects of psychosocial changes.
- Describe essential aspects of cognitive changes.
- Explain essential aspects of moral development.
- Explain essential aspects of spiritual development.
- Identify common health concerns and hazards of older adults.
- Discuss nursing implications of the common health concerns and hazards identified.

DEVELOPMENTAL ASPECTS OF LATE ADULTHOOD

Briefly describe the physiologic, psychosocial, and cognitive aspects of late adulthood development in the table below.

1. PHYSIOLOGIC DEVELOPMENT	
2. PSYCHOSOCIAL DEVELOPMENT	
3. COGNITIVE DEVELOPMENT	

Briefly describe the developmental tasks and the spiritual and moral aspects of late adulthood in the table below.

4. DEVELOPMENTAL TASK	
5. MORAL DEVELOPMENT	
6. SPIRITUAL DEVELOPMENT	

NURSING IMPLICATIONS FOR HEALTH PROMOTION

Identify three health concerns or hazards that commonly occur in late adulthood, then discuss the related nursing implications in the table below.

HEALTH CONCERN OR HAZARD	NURSING IMPLICATIONS
7.	
8.	
9.	

SELF ASSESSMENT QUESTIONS

10. When feeding an elderly person the nurse should

    A. increase calories.
    B. offer smaller servings.
    C. reduce fluid.
    D. increase fat.

11. Older people thrive on being independent because doing tasks for themselves

    A. costs less.
    B. fosters self-respect.
    C. maintains privacy.
    D. encourages socialization.

12. A major concern in providing nursing care for the elderly is to

    A. assist with financial planning.
    B. enhance family relations.
    C. foster independence.
    D. provide a religious environment.

13. Life expectancy of North Americans is

    A. remaining constant.
    B. increasing.
    C. decreasing.
    D. unknown.

14. As a person ages, his or her intellectual abilities

    A. are unaffected.
    B. increase.
    C. change minimally.
    D. decrease.

ADDITIONAL LEARNING ACTIVITIES

1. Interview a young adult, a middle-aged adult and an elderly person. Compare their developmental needs and problems. How are they alike? How do they differ?

2. Interview and assess an elderly person in your family. Compare his or her growth and development to the standards outlined in this chapter for physical, psychosocial, cognitive, spiritual and moral development.

3. Prepare a teaching plan about nutrition, safety or personal hygiene for the elderly. Teach this information to an elderly family member or friend.

28 FAMILY HEALTH

Throughout the ages, nursing has been concerned with the health needs of the individual. But as nursing developed holistic, systems, and developmental approaches to client care, the need for including the family in the nursing process became apparent. The nursing profession recognized that the influence the family had on an individual's response to disease and injury was significant. The nurse's understanding of the family's social interaction and communication patterns became just as important as the psychosocial, physiologic, cultural, and spiritual elements of providing client care. Today the professional nurse uses a variety of approaches that incorporate the role of the family when applying the nursing process to client care. In this capacity, the nurse is responsible for meeting the needs of the family as a whole as well as each family member. This chapter focuses on the knowledge and skills the nurse must possess to apply the nursing process to the family and its members. After completing this chapter the student will be able to:

- Explain family-centered nursing.
- Describe different kinds of families.
- Describe the roles and functions of the family.
- Identify three frameworks for studying the family.
- Describe selected forms of families in society today.
- Identify the components of a family health assessment.
- Identify common risk factors to family health.
- Develop nursing diagnoses pertaining to family functioning.
- Develop outcome criteria for specific nursing diagnoses related to family functioning.
- Explain the common causes of a family health crisis.
- Discuss coping strategies used by families.
- Explain the nurse's function in family health promotion.

ROLES AND FUNCTIONS OF THE FAMILY

Define the following types of families.

1. Nuclear family ______________________________

2. Extended family ______________________________

3. Traditional family ______________________________

4. Blended family ____

5. Communal family ____

ASSESSING THE HEALTH OF FAMILIES

Case Study:

Jim Washington is a 36-year-old black male who sells computers for a large company with branches throughout the country. His frequent travel often keeps him away from home. His major health problems are hypertension, slight obesity, and stress-related peptic ulcers. Luwanda, his 35-year-old wife, recently began work as an accountant for a local trucking firm. She works long hours but enjoys the extra income for her family that the job provides. Her major health problems are obesity and borderline diabetes. Jonella, the 12-year-old daughter of the Washingtons, is a 7th-grade student in a school near their home. Jonella has been diagnosed with sickle cell anemia and misses school frequently because of frequent colds and joint pain. She expresses dissatisfaction with her family's life-style, stating, "Mom and Dad are gone too much! I don't like being a latchkey kid! I hate to be home alone when I am sick!" Janetta, the Washingtons' oldest daughter, is 14 years old and a freshman at the local high school. Very active, she spends a great deal of her time in extracurricular activities at school or visiting friends. She says, "It doesn't bother me that Mom and Dad are gone so much. I've got my own life to lead! I don't stay home that much because Mom and Dad are never around and my sister is a drag!" Jim and Luwanda do not appear to be disturbed by the girls' remarks. "They're always complaining about something," Jim says.

There is a clear sense of warmth and affection of the parents for their children but in all family-related issues Jim is domineering, authoritative, and makes all the decisions concerning the family and its members. Although the youngest daughter sends clear messages of distress and loneliness, the other family members appear not to listen. The family attends church every Sunday but shares no other activities.

Complete the following family assessment for the Washington family.

ASSESSMENT DATA	JIM	LUWANDA	JONELLA	JANETTA
6. Age				
7. Occupation				
8. Stressors				
9. Health problems				

Complete the following APGAR assessment information about the Washington family.

COMPONENT	WASHINGTON FAMILY DATA
A DAPTATION	10.
P ARTNERSHIP	11.
G ROWTH	12.
A FFECTION	13.
R ESOLVE	14.

Identify three risk factors in the Washington family.

15. ____________________

16. ____________________

17. ____________________

DIAGNOSING AND PLANNING

Develop two nursing diagnoses and two related outcome criteria for the Washington family.

NURSING DIAGNOSES	OUTCOME CRITERIA
18.	
19.	

THE FAMILY EXPERIENCING A HEALTH CRISIS

Discuss the factors that help a family cope with the illness of a family member.

20. ____________________

____

____

SELF ASSESSMENT QUESTIONS

21. The Washington family can be described as

    A. traditional.
    B. blended.
    C. reconstituted.
    D. extended.

22. The statement that is *least* likely to be used as a nursing diagnosis for the Washington family is

    A. ineffective grieving
    B. alteration in parenting.
    C. impaired verbal communication.
    D. ineffective family coping.

23. When assessing the Washington family's communication patterns, the nurse should pay the closest attention to

    A. where family members sit during the discusssion.
    B. how well the children mind their parents.
    C. how long the discussion lasts.
    D. who does most of the talking.

24. Family centered nursing is

    A. caring for traditional families.
    B. considering the health of the family as a unit.
    C. functioning in a family clinic.
    D. functioning as a family nurse practitioner.

25. The structural-functional theory for studying the family focuses on the family's

    A. growth and development.
    B. critical tasks.
    C. hierarchy of needs.
    D. stucture.

ADDITIONAL LEARNING ACTIVITIES

1. Interview a family in the community. Complete a health appraisal on the family.

2. Develop a nursing care plan for the family you assess in activity 1.

3. Plan a one-hour visit with a traditional, a blended, and a single-parent family. Compare and contrast the following:
   a. Communication patterns
   b. Decision making
   c. Discipline

29 SELF-CONCEPT

The ideal the client develops about him or herself and the ability to accomplish adequate role relationships is developed throughout a lifetime of relating to others. As a small child an individual takes the reactions received from significant others, integrates this feedback with his or her own self-perception, and through this process develops feelings and ideas about the self. This chapter focuses on processes underlying the development of self-concept and its effect on role relationships. After completing this chapter the student will be able to:

- Differentiate self-concept from self-esteem.
- Describe the components of self-concept.
- Give Erikson's explanation of the effects of psychosocial crises on self-concept and self-esteem.
- Describe the effects of communication/coping styles on self-esteem.
- Identify four areas involved in the nursing assessment of self-concept.
- Describe key data to be included when assessing self-perception.
- Describe the essential aspects of assessing roles and relationships.
- List important assessment data to be included when identifying clients' stressors and coping strategies.
- Identify common stressors affecting self-concept and self-esteem.
- List behaviors that could indicated altered self-concept.
- Identify nursing diagnoses concerning altered self-concept.
- Select appropriate goals for clients with altered self-concept.
- Describe nursing actions designed to implement identified goals for clients with altered self-concept.
- Describe ways to enhance the self-esteem of older adults.
- Identify outcome criteria that permit evaluation of clients with altered self-concept.

CONCEPT OF SELF AND SELF-ESTEEM

Describe the difference between self-concept and self-esteem.

1. ____

____

List the four components of the self-concept.

2. ______________________________

3. ______________________________

4. ______________________________

5. ______________________________

Case Study:

John Greenstein, a 20-year-old construction worker, was confused, frightened, and frustrated. He had fallen beneath the wheels of an earthmoving machine at work and his leg had been amputated. During a clinical conferance John's primary nurse discussed his progress. "John is having a great deal of difficulty adjusting," she said. "He refuses to talk to us and gets angry when we try to involve him in any activities." She noted that John's mother had mentioned that she and her husband had adopted John at four years old after he had been abused and abandoned by his natural parents.

Does John's early background impact on how he is handling this crisis in his life? If so, describe how.

6. ______________________________

____

____

____

The nurses notice that John tends to be aggressive rather than assertive when coping with his condition. What is the difference between aggressive and assertive behavior?

7. ______________________________

____

____

DEVELOPMENT OF SELF-ESTEEM

Describe the four elements that influenced the development of John's self-esteem in the space provided.

8.	10.
9.	11.

ASSESSING

John continued to be angry, anxious, and depressed. He seldom talked to the unit nurses and remained isolated in his room. He seemed to have difficulty adjusting to the demands of physical therapy. Even though he made excellent progress with exercises and crutchwalking, he continued to complain of being a failure. One night while changing John's stump dressings the night nurse noticed that he refused to look at the incision. He said, "I'm not surprised this happened to me. I think I have a black cloud over my head." His nurse noticed that John often made pessimistic comments like this.

12. John's comments are a reflection of his ______________________________

John's nurse believes that he has a distorted self-perception. What observations should she make to assess John's self-perception.

13. ____

____

Identify and give an example of two types of irrational, illogical thinking that John's comments reflect.

14. ____

____

15. ____

____

John's steady girlfriend, Gina, is disturbed by John's reaction to her. One evening the nurse finds Gina crying in the visitor's lounge. "I know John is going through a lot," she says, "but I can't understand why he hates me. What have I done? Last night he told me to find another boyfriend. He said I have a good excuse to dump him." Gina pleads with the nurse, "Please tell me why John is saying these things. If I understood, perhaps I could help."

If you were in the nurse's place, what would you tell Gina?

16. ____

____

____

John, like many clients who experience a serious change in body image, is having difficulty in understanding how his amputation will affect his relationship with Gina. What kind of information should the nurse obtain when assessing the client's role and relationship with his family and friends?

17. ____

____

____

John continued to be upset, angry, and depressed. John's nurse was anxious to help him adjust to the changes in his life. One morning while helping him get ready for physical therapy she asked, "Has anything as serious as this happened to you before?" John told her that the only time he had felt this bad was when his best friend was killed in a car accident when he was 14 years old. The nurse asked John to tell her how he had coped when his friend died.

Why is the nurse asking John these questions?

18. ____

____

DIAGNOSING, PLANNING IMPLEMENTING AND EVALUATING

Complete the care plan for John below by completing the two nursing diagnoses. For each nursing diagnosis develop one client goal and two outcome criteria. Develop at least two nursing interventions and related rationales for each diagnosis.

NURSING DIAGNOSIS	CLIENT GOALS/ OUTCOME CRITERIA	NURSING INTERVENTIONS AND RATIONALES
19. Body Image disturbance related to:	20.	21.
22. Ineffective individual coping related to:	23.	24.

SELF ASSESSMENT QUESTIONS

25. John frequently makes negative comments to the staff, such as, "Nobody cares about me."

    A. "John doesn't like being in the hospital.
    B. "John is frustrated because he can't go home. Let's humor him and keep him busy.
    C. "John is angry and depressed about his situation. Let's help him identify the cause of his anger."
    D. "John is trying to get our attention. Let's ignore his behavior."

26. During a conference, the staff identifies one of John's problems as a feeling of powerlessness during this hospitalization. Which of the following plans of action would be most helpful in dealing with this problem?"

    A. Ask John to call you by your first name; spend more time with him.
    B. Ask John to be more cooperative. Give him a schedule to follow.
    C. Tell John he's in charge. Follow his orders regarding his care.
    D. Respect John's privacy. Ask for his input in his plan of care.

27. John is having difficulty adjusting to his new body image. He seems most self-conscious when his girlfriend is present. This most likely happens because

    A. John's distorted body image is affecting his sexual self-concept and he may feel sexually inadequate.
    B. John's self-concept has been lowered by the amputation and he does not feel his girlfriend can respect him.
    C. John may believe that she will be "turned off" by his disfigurement and that she may want to end the relationship.
    D. John needs time to adjust to his change in self-image before he can relate to a woman again.

28. John refuses to participate in the care of his stump. Which of the following actions by the nurse would help John accept his body-image change?

    A. Focus on his body strengths rather than his weaknesses.
    B. Tell John he's making himself more sick.
    C. Focus on the surgery. Encourage him to talk about the accident.
    D. Insist that he look at the stump.

ADDITIONAL LEARNING ACTIVITIES

1. Observe a client who is having difficulty with self-concept. Identify behaviors the client displays that support your assessment. Develop a nursing diagnosis related to self-concept for the client.

2. Interview an elderly person in a long-term care agency. Assess how this person perceives and values himself or herself.

3. Interview several pregnant women in various stages of pregnancy. Compare how each one feels about her body image.

4. In a clinical area, select a client who has demonstrated one of the following behavior problems: anger, isolation or help-seeking behavior.

30 SEXUALITY

An understanding of human sexuality is essential for the nurse providing holistic nursing care to an individual throughout the lifespan. Along with understanding, the nurse must also develop a sense of self-awareness so that her or his personal biases do not interfere with the sexual value system of the client. After completing this chapter the student will be able to:

- Describe selected aspects of sexuality.
- Identify key aspects of the development of sexuality from the prenatal period to late adulthood.
- Compare selected physical and psychologic sexual stimulation patterns.
- Identify physiologic changes occuring in males and females during each phase of the sexual response as described by Masters and Johnson.
- List factors that affect an individual's sexual attitudes and behaviors.
- Give examples of how to obtain data about sexual functioning when conducting a health history.
- Identify factors contributing to sexual dysfunction.
- List factors that increase and decrease sexual motivation.
- Describe common problems of genital sexuality and possible causes.
- Identify common illnesses affecting sexuality.
- Compare selected intervention models for sexual counseling.
- Describe key points about breast and testicular self-examinations to include in health teaching.
- Identify essential aspects about selected contraceptive methods to include in health teaching.
- Describe guidelines for the prevention of sexually transmitted diseases.
- Describe essential outcome criteria that permit evaluation of client progress toward meeting planned goals.

SEX AND SEXUALITY

List three reasons sexuality is a more complex issue for nurses practicing today than it has been in the past.

1. ______________________________

2. ______________________________

3. ____

How do historical, ethnocultural, religious and contemporary factors influence sexuality?

4. Historical ____

____

5. Ethnocultural ____

____

6. Religious ____

____

7. Contemporary ____

____

DEVELOPMENT OF SEXUALITY

Discuss how sexuality develops for each period of growth and development.

GROWTH AND DEVELOPMENT	BIOLOGIC	PSYCHOSOCIAL
8. Infancy		
9. Childhood		
10. Adolescence		
11. Adulthood		
12. Late adulthood		

PATTERNS OF SEXUAL FUNCTIONING

Briefly describe the physiologic changes that occur in males and females during each phase of sexual response .

FEMALES	MALES
13. Excitement phase	17. Excitement phase
14. Plateau phase	18. Plateau phase
15. Orgasmic phase	19. Orgasmic phase
16. Resolution phase	20. Resolution phase

ASSESSING SEXUAL HEALTH

Discuss how information about a client's sexual health should be obtained. Give several examples of questions that could be used.

21. ____

____

____

____

List at least 5 factors that contribute to sexual dysfunction.

22. ____

23. ____

24. ____

25. ____

26. ____

List two factors that decrease sexual motivation and two factors that decrease sexual motivation.

27. ____________________ 29. ____________________

28. ____________________ 30. ____________________

Identify the following terms in column I by drawing a line to the appropriate description in colum II.

31. Erectile dysfunction
32. Premature ejaculation
33. Primary orgasmic dysfunction
34. Situational orgasmic dysfunction
35. Vaginismus
36. Dyspareunia

A. Inability of the male to delay ejaculation
B. Pain experienced by a women during intercourse
C. A woman who has never been able to achieve orgasm
D. The inability to achieve or maintain an erection
E. The irregular and involuntary contraction of the muscles around the outer 1/3 of the vagina during coitus
F. A woman who has previously been able to experience orgasm, but is unable to do so in the present

DIAGNOSING AND PLANNING SEXUALITY PROBLEMS

Develop two nursing diagnoses and two related outcome criteria for clients with sexual/sexuality problems.

NURSING DIAGNOSIS	OUTCOME CRITERIA
37.	
38.	

IMPLEMENTING

List at least six strategies the nurse should use to prevent the spread of AIDS.

39. ______________________________

40. ______________________________

41. ______________________________

42. ______________________________

43. ______________________________

44. ______________________________

Describe each of the following contraceptive methods.

45. Coitus interruptus______________________________

____

46. Condom ______________________________

____

47. Vaginal diaphragm______________________________

____

48. Oral contraception____

____

49. Tubal ligations ____

____

50. Vasectomy ____

____

SELF ASSESSMENT QUESTIONS

51. Sexual behavioral differences are not apparent between males and females during infancy and childhood. This statement is

    A. true.
    B. false.

52. Adolescents have problems with

    A. sexual self-identity.
    B. gender identity.
    C. role identity.
    D. biologic identity.

53. The sex of a fetus is determined by

    A. the XX chromosomes of the female.
    B. the XY chromosomes of the male.
    C. mitosis
    D. meiosis

54. The sexual response cycle includes the

    A. excitement, plateau, orgasmic, and resolutions phases.
    B. excitement, orgasmic, resolution, and plateau phases.
    C. plateau, excitement, orgasmic, and resolution phases.
    D. orgasmic, excitement, resolution, and plateau phases.

55. The characteristics of positive sexual health in an individual include

    A. an expression of a positive body image.
    B. the ability to create effective relationships with both sexes
    C. the acceptance of responsibility for pleasure and reproduction.
    D. all of the above.

ADDITIONAL LEARNING ACTIVITIES

1. Observe a group of children at play. Identify those activities related to sex role and sexual identify.

2. Interview parents of a young child. Assess how they contribute to the development of the child's identity and role.

3. Interview a physician, nurse practitioner, or both in the community. Determine what sexual problems she or he encounters in practice.

4. Discuss with an adolescent what she or he wants to know about sex. Find out where the adolescent obtains information about sex.

5. Interview an individual who is homosexual. What kinds of problems does this person experience living in today's society? What special concerns does he or she have about being a client in a hospital?

6. Assess your own thoughts, feelings and biases concerning sexuality issues. Discuss your thoughts and feelings with a group of your contemporaries.

31 ETHNIC AND CULTURAL VALUES

The melting pot of nationalities that has immigrated to North America from around the world has created a challenge for the nurse who strives to integrate a client's ethnic and cultural background into the plan of care. In order to do this the nurse needs a broad knowledge and sensitivity to the ethnic, racial, and cultural uniqueness of each client. Chapter 31 focuses on the knowledge and skills the nurse must possess to incorporate these ethnic and cultural aspects into the nursing process. After completing this chapter the student will be able to:

- Describe the concept of culture.
- Identify characteristics and universal attributes of culture.
- Identify social characteristics common to all ethnic/cultural groups that health care providers must consider.
- Identify problems unique to ethnic minorities in the provision and use of health care services.
- Relate the incidence of specific diseases to certain ethnic or cultural groups.
- Identify specific characteristics and values of selected cultural groups that may influence nursing assessment and intervention.
- Relate health-related beliefs and practices to economic status.
- Contrast the values of the health care culture and selected minority ethnic cultures.

CONCEPTS OF ETHNICITY AND CULTURE

What is the difference between the following sets of terms?

Ethnic group and race:

1. ____

____

Material and nonmaterial culture:

2. ____

____

Dominant and minority groups:

3. ____

____

Match each of the terms in column I with the appropriate term in column II.

4. _____ Ethnicity
5. _____ Race
6. _____ Culture
7. _____ Dominant group
8. _____ Acculturation
9. _____ Ethnocentrism
10. _____ Stereotyping
11. _____ Racism
12. _____ Ethnoscience
13. _____ Culture universals

A. The beliefs and practices that are shared by people and passed down from generation to generation

B. Cultural assimilation

C. Assuming that all members of a culture or ethnic group are alike

D. The condition of belonging to a specific ethnic group

E. The common features of behavior that are similar among different cultures

F. A system of classifying humans into subgroups according to physical characteristics

G. Provide the controlling value system in society

H. The belief that one's own culture is superior to all others

I. The systematic study of the way of life of a designated cultural group

J. The assumption of racial superiority or inferiority

List the seven characteristics of culture in the space provided.

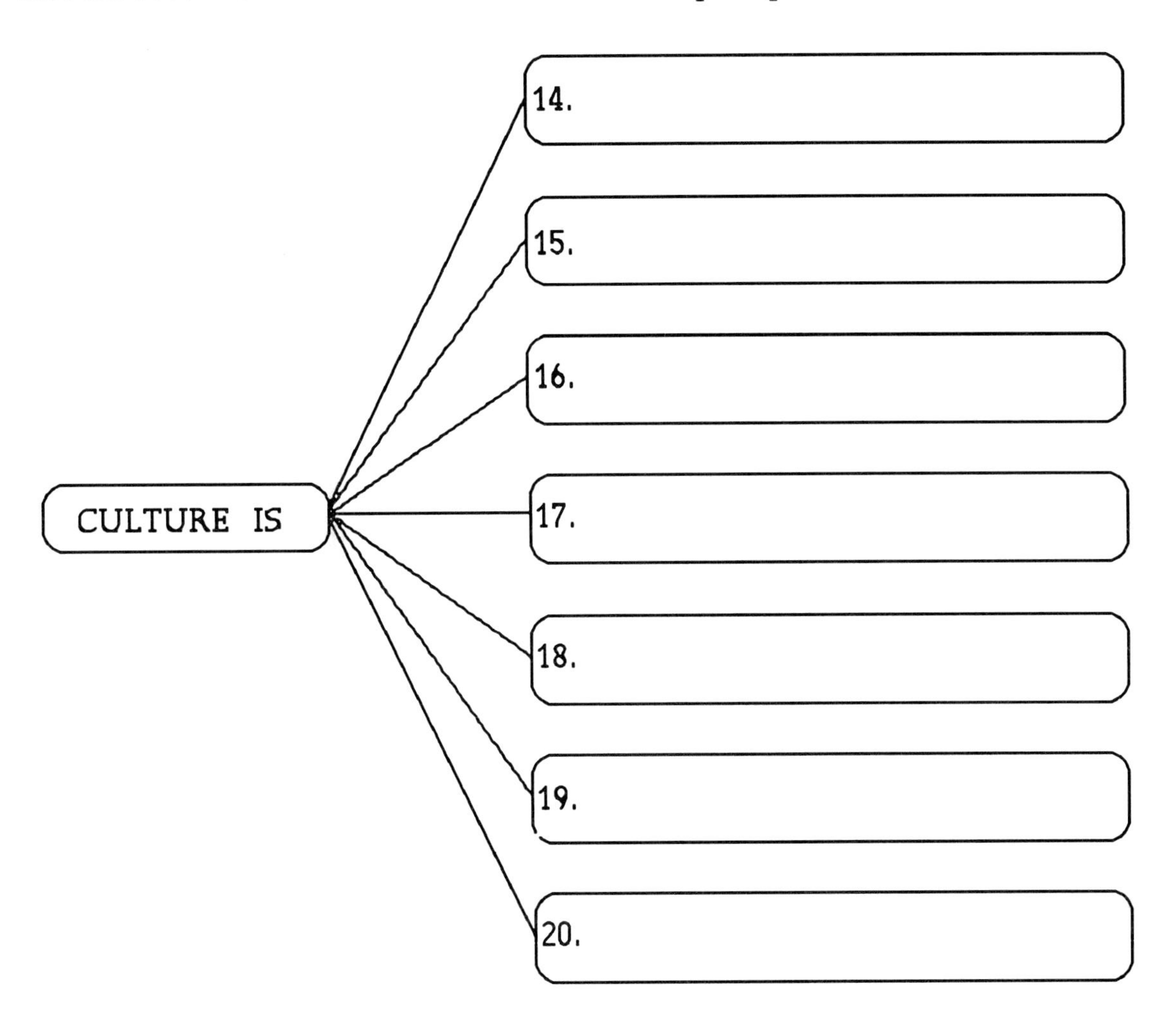

DIVERSITY OF NORTH AMERICAN SOCIETY

What does the nurse need to know about each of the following areas in order to meet the ethnocultural needs of the client?

Male-female roles:

21. ____

Language and communication patterns:

22. ____

Territoriality and personal space:

23. ______________________________

Time orientation:

24. ______________________________

Family:

25. ______________________________

Food and nutritional practices:

26. ______________________________

SUSCEPTIBILITY TO DISEASE

Which group(s) is/are more susceptible to the following diseases?

27. Sickel-cell anemia ______________________________

28. Hypertension ______________________________

29. Diabetes mellitus______________________________

30. Cancers ______________________________

31. Alcohol metabolism problems______________________________

FOLK HEALING AND TRADITIONAL WESTERN MEDICINE

Why do individuals tend to use nontraditional folk healing methods rather than the generally accepted treatments in the dominant society?

32. ______________________________

____

____

CULTURAL AND ETHNIC GROUPS IN NORTH AMERICA

Fill in the box in the game below by placing the cultural/ethnic group identified below in the appropriate space.

	E	T	H	N	O
1	33.	38.	43.	48.	53.
2	34.	39.	44.	49.	54.
3	35.	40.	45.	50.	55.
4	36.	41.	46.	51.	56.
5	37.	42.	47.	52.	57.

E1 Not concerned about future
E2 Includes Chinese, Japanese, Koreans, and Vietnamese
E3 Recent immigrants come from Jamiaica, and Haiti
E4 Believe in "hot" and "cold" or "wet" and "dry" theories
E5 May show defensive and hostile behaviors if they are not familiar with the health care system
T1 Leadership positions given to the elderly
T2 Includes people from Egypt, Lebanon, Saudi Arabia, Kuwait among others
T3 Thalasemia is a major health problem
T4 Folk healer is called a curandero
T5 Believe direct eye contact takes anothers soul away
H1 Strongly emphasize harmony and avoidance of conflict in groups
H2 May view health care professionals as their employees
H3 Use witchcraft and voodoo.
H4 Drug addiction is a major health problem
H5. Follow the "ethic of neutrality"
N1 Soul food is the traditional diet
N2 Each person speaks only for him or herself
N3 May reclaim amputated limbs for burial
N4 Low protein and iron in diet.
N5 Health is being in total harmony with nature
O1 Major health problem are hypertension, sickle-cell
O2 Experience Mal de ojo
O3 Avoid attracting special attention to themselves
O4 Leading causes of death are accidents, suicide, diabetes, alcoholism, and homicide
O5 Matriarchal family system

HEALTH CARE SYSTEM AS A SUBCULTURE

Complete the following sentence:

58. The health care system should be considered a subculture because ____________

____

CULTURE SHOCK

Describe the four phases of culture shock.

59. ____

60. ____

61. ____

62. ____

APPLYING THE NURSING PROCESS

Develop two nursing diagnoses and two related outcome criteria for the client who has special cultural/ethnic needs.

NURSING DIAGNOSIS	OUTCOME CRITERIA
63.	
64.	

SELF ASSESSMENT QUESTIONS

65. Which of the following statements is untrue? An ethnic group always shares the same

    A. racial group.
    B. cultural heritage.
    C. social heritage.
    D. folktales.

66. Culture is

    A. learned through life experiences after birth.
    B. developed through the interactions of people.
    C. transmitted from parents to children.
    D. all of the above.
    E. none of the above.

67. The nurse needs to know who the dominant person in the family is because the dominant person

    A. is the health care provider for the family.
    B. tends to make health care decisions.
    C. determines the family value system regarding health.
    D. is the only one who will communicate with the nurse.

68. Keloids are most commonly seen in

    A. Black Americans.
    B. Asian Americans.
    C. Hispanic Americans.
    D. Appalachian Americans.

69. When assessing a client's culture, the nurse should consider

    A. the client's value system.
    B. the client's belief system.
    C. the client's customs.
    D. all of the above.
    E. none of the above.

ADDITIONAL LEARNING ACTIVITIES

1. Select a client with a different ethnic and cultural background than your own.
   A. Attempt to learn more about his or her unique customs and beliefs.
   B. Determine how the nursing care plan could be adjusted or accommodated to his or her differences.

2. Discuss your attitudes and beliefs concerning folk medicine, faith healers, and spiritualism with a group of your colleagues.

32 SPIRITUAL AND RELIGIOUS BELIEFS

Down through the ages nurses have recognized that clients who had an awareness of their own spirituality were more able to cope with the effects of illness. And, if the client were able to actualize his or her spirituality into a religious belief system, the ability to cope with illness became even stronger. One of the major responsibilities for nursing then, is to be able to incorporate the client's spiritual and religious beliefs into the plan of care. In order to do this the nurse must become adept at assessing and diagnosing spiritual distress in the client. Nurses must also be aware of, respect, and incorporate into the plan of care, that some clients do not believe in the spiritual nature in man or the existence of a higher being. After completing this chapter the student will be able to:

- List essential facts about spiritual beliefs and religious practices as they relate to health care.
- Describe essential facts about the spiritual development of different age groups.
- List expressions of spiritual well-being.
- Identify categories of information to obtain in the nursing assessment through nursing history and clinical assessment.
- List essential aspects of nursing diagnosis related to spiritual care.
- Explain facts about nursing interventions to support clients' spiritual beliefs and religious practices.
- State outcome criteria essential for evaluating the clients' progress.

SPIRITUALITY, FAITH, AND RELIGION

Define the following terms.

1. Spirituality ______________________________

____

2. Faith ______________________________

____

3. Religion ______________________________

____

RELIGION AND ILLNESS AND SPIRITUAL DEVELOPMENT

True or False

4. ______ Spiritual beliefs influence life-style, attitudes, and feelings about illness and death.

5. ______ Toddlers have a beginning sense of right and wrong.

6. ______ Children learn what is considered good and bad from parents.

7. ______ Five-year-olds have no understanding of God.

8. ______ Seventh Day Adventists urge members not to take drugs unless absolutely necessary.

9. ______ School-age children begin to understand that prayers are not always answered.

10. ______ By 12 years, many children have decided whether to accept the family religion.

11. ______ Most adolescents are unable to have faith or accept a religious belief system.

12. ______ When people associate disease with immoral behavior they may believe their illness is punishment for past sins.

13. ______ Middle-aged adults tend to become more involved in religious activities.

14. ______ Elderly adults who do not have mature religious beliefs may experience a feeling of deprivation as they become less active.

SPIRITUAL HEALTH AND THE NURSING PROCESS

Develop two nursing diagnoses and two related outcome criteria for the client who has spiritual distress.

NURSING DIAGNOSIS	OUTCOME CRITERIA
15.	
16.	

RELIGIOUS BELIEFS RELATED TO HEALTH CARE

For each of the following religions, describe at least one nursing implication or intervention the nurse must employ when caring for a client of that faith.

RELIGION	NURSING INTERVENTION
CHRISTIAN SCIENTIST	17.
BAPTIST	18.
JEHOVAH'S WITNESS	19.
JUDAISM	20.
MUSLIM/MOSLEM (ISLAM)	21.
ROMAN CATHOLICISM	22.

SELF ASSESSMENT QUESTIONS

23. An agnostic is a person who

    A. denies the existence of God.
    B. believes in the existence of one God.
    C. believes in the existence of several gods.
    D. doubts the existence of God.

24. Faith is

    A. a belief in a higher power.
    B. a person's need to fulfill religeous obligations.
    C. an organized system of worship.
    D. to believe in or be committed to something or someone.

25. In Christian Science, the role of ministering to the sick is carried out by a

    A. practitioner.
    B. rabbi.
    C. priest.
    D. minister.

26. The doctrine of avoidance of extremes is practiced by

    A. Christian Scientists.
    B. Buddhists.
    C. Mormons.
    D. Seventh Day Adventists.

27. Orthodox and Conservative Jews

    A. are baptized at birth.
    B. have no dietary restrictions.
    C. observe as the Sabbath from sunset Friday to sunset Saturday.
    D. require abstinence from alcohol.

ADDITIONAL LEARNING ACTIVITIES

1. Interview a member of the pastoral care department of a local hospital. Ascertain how hospital chaplains meet the spiritual needs of a client in spiritual distress.

2. Interview a nurse who cares for hospice clients. Determine how he or she meets the spiritual needs of clients who are dying.

33 STRESS TOLERANCE, AND COPING

A major focus of nursing is to assist the client maintain physiological, psychological, and sociological equilibrium, a task that requires a broad knowledge of the dynamics of homeostasis and stress. Chapter 33 offers an indepth discussion of these areas so that the student will be able to:

- Give four main characteristics of homeostatic mechanisms.
- Explain how the autonomic nervous system and the endocrine system regulate homeostasis.
- Describe how the respiratory, cardiovascular, renal, and gastrointestinal systems interact to maintain homeostasis.
- Differentiate the concepts of stress as a stimulus, as a response, and as a transaction.
- Identify Selye's definition of stress.
- Describe the three stages of Selye's general adaptation syndrome.
- Describe essential aspects of the Lazarus stress model.
- Identify physiologic and psychologic (cognitive, verbal, and motor) manifestations of stress.
- Identify behaviors related to specific ego defense mechanisms.
- Differentiate four levels of anxiety.
- Give examples of constructive and destructive anger.
- Give examples of three modes of adaptation.
- Identify examples of nursing diagnoses related to stress.
- Identify general guidelines to minimize a client's anxiety and stress.
- Identify interventions to help clients cope with stress.
- Describe outcome criteria that can be used to evaluate whether a client is effectively coping with a stressful problem.
- Identify sources of stress in the nurse.

HOMEOSTASIS

The four main characteristics of homeostatic mechanisms are:

1. ______________________________

2. ______________________________

3. ______________________________

4. ____

Identify the hormones and the functions produced by the organs of the endocrine system.

ORGAN	HORMONES	FUNCTIONS
5. Anterior pituitary		
6. Posterior pituitary		
7. Adrenal medulla		
8. Adrenal cortex		
9. Thyroid		
10. Para-thyroids		
11. Isle of Langerhans		

Describe the functions of the respiratory, cardiovascular, renal, and gastrointestinal systems in maintaining homeostasis.

SYSTEM	FUNCTION
12. Respiratory	
13. Cardiovascular	
14. Renal	
15. Gastrointestinal	

CONCEPT OF STRESS

Identify and define the three concepts of stress.

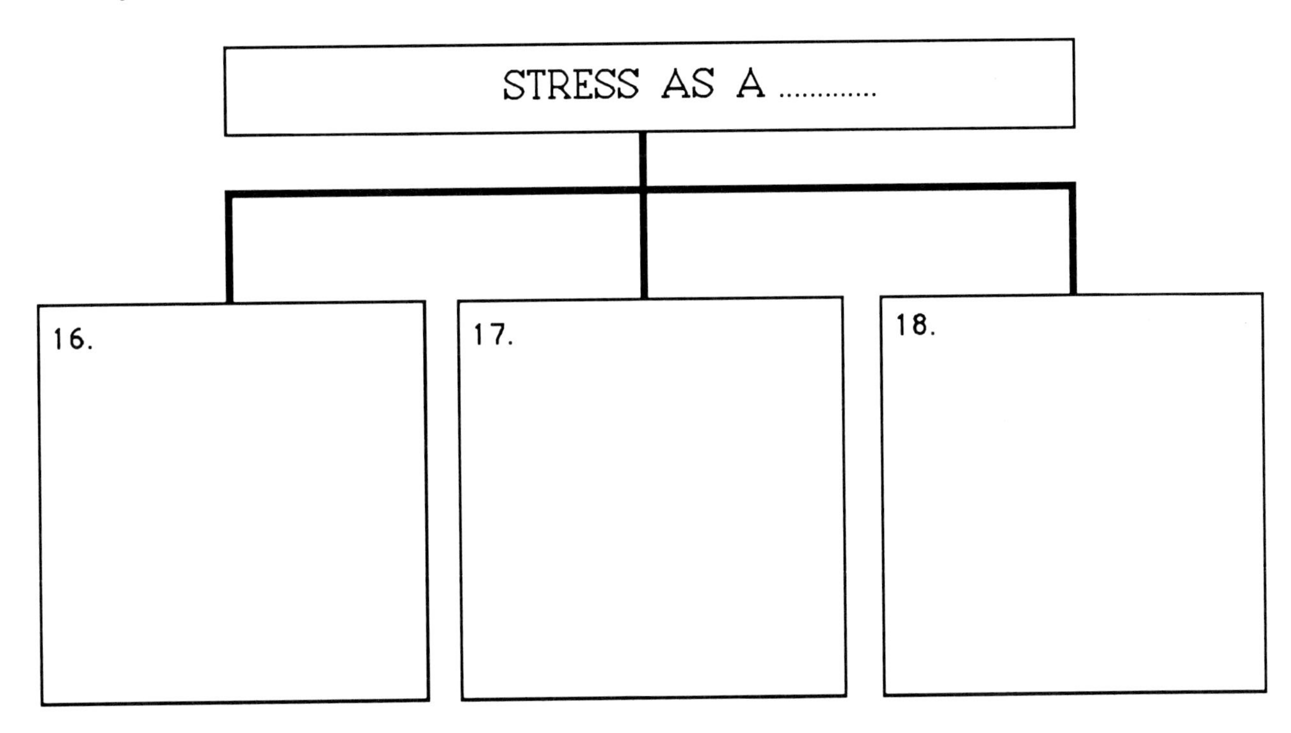

Define stress according the Selye.

19. ____

____

Identify and describe the three stages of Selye's general adaptation syndrome.

20. ____

____

21. ____

____

22. ____

____

MANIFESTATIONS OF STRESS

Define the following manifestations of stress.

23. Anxiety ____

____

24. Anger ____

____

25. Suppression ____

____

26. Defense mechanisms ____

____

Describe the four levels of anxiety.

27. LEVEL I

28. LEVEL II

29. LEVEL III

30. LEVEL IV

ADAPTATION

Differentiate between the physiologic and psychologic modes of adaptation to stress.

31. ____

____

____

USING THE NURSING PROCESSS

Develop two nursing diagnoses that would be appropriate for a client who is dealing with stress. Then develop a related client goal/outcome criteria and nursing intervention for each nursing diagnosis in the table below.

NURSING DIAGNOSIS	CLIENT GOAL/ OUTCOME CRITERIA	NURSING INTERVENTION AND RATIONALE
32.		
33.		

STRESS MANAGEMENT FOR NURSES

List at least three techniques the nurse can use to minimize his or her own anxiety.

34. ____________________

35. ____________________

36. ____________________

SELF ASSESSMENT QUESTIONS

37. The antidiuretic hormone (ADH), is sometimes referred to as a vasopressor because it can

    A. induce sodium chloride retention.
    B. metabolize glucose.
    C. elevate blood pressure.
    D. control growth rate.

38. Hormones that control fluid and electrolytes levels are produced by

    A. the anterior pituitary gland.
    B. the posterior pituitary gland.
    C. the adrenal medulla.
    D. the adrenal cortex.

39. In order to minimize their own stress nurses should

    A. ignore feelings of anxiety.
    B. refrain from sharing feelings.
    C. take defensive measures.
    D. accept what cannot be changed.

40. Redirecting libidinal drives into socially acceptable channels is a description of

    A. projection.
    B. sublimation.
    C. displacement.
    D. reaction formation.

41. An increased capacity of the heart and lungs after prolonged exercise is a(n)

    A. physiologic mode of adaptation.
    B. psychologic mode of adaptation.
    C. sociocultural mode of adaptation.
    D. environmental mode of adaptation.

ADDITIONAL LEARNING ACTIVITIES

1. Develop a list of stress producing situations you have experienced within the past three months.
2. Develop a strategy for reducing stress for each of the items you listed in the first activity.
3. Select a friend or client who seems to be undergoing physical, psychologic or sociologic stress. What defense mechanisms is he or she demonstrating?
4. Interview a nurse who works in a stress producing clinical area. What does this nurse do to relieve stress?

34 COPING WITH LOSS, GRIEVING, AND DEATH

At the beginning of this century the experience of dying and death was a family affair with the dying person being cared for by family, friends and neighbors at home. With the advent of modern technological advances however, the care of the dying moved from the home to the hospital, and before long more than 80% of all clients died in the hospital away from the support systems that they needed so badly. Their families also suffered; unable to experience the dying process at close hand, they found the loss difficult to accept and their grief was prolonged. Today, nurses have become more aware of the importance of assessing the client and family's response loss, grieving and dying. This chapter provides the nurse with a framework for using the nursing process in assisting the client and family deal with loss, grieving and dying. After completing this chapter the student will be able to:

- Recognize selected frameworks for identifying stages of grieving.
- Identify clinical symptoms of grief.
- Discuss factors affecting a loss reaction.
- Recognize common fears associated with dying.
- Identify factors contributing to unresolved grief.
- Describe guidelines for helping the client to die with dignity.
- Identify measures that facilitate the grieving process.
- List changes that occur in the body after death.
- Describe nursing measures for care of the body after death.

LOSS AND GRIEF

Mary Jefferson put her pen down as she thought about Ray Morris, the client she had cared for during the last few days. Ray was the first dying client she had nursed since graduation and, although it was difficult, she found that writing a journal helped her sort out her feelings. Mary had lost her father six months ago and caring for Ray had reawakened painful memories. She still had difficulty believing that her father was really gone. "Sometimes I start dialing the telephone before I remember he's not there." Mary began to write....

June 23:
"Nurse, am I dying?" he looked in my eyes with such trust, I knew I could not lie. Even though his family and doctor had agreed not to tell him his prognosis, he knew that he was dying. I struggled with my ethics, values and conscience for just one moment. Then I

said, "Yes Ray, you and I both know how ill you are. What can I do to make it easier?" I stroked his forehead and held his hand. He smiled, closed his eyes and slept, the first peaceful time he had experienced in days.

Mary is still grieving for her father. Describe Martocchio's five clusters of grief that family members experience.

1. ______________________________

2. ______________________________

3. ______________________________

4. ______________________________

5. ______________________________

Mary's father died after a long-term illness, so she had the experience of going through anticipatory grief. What is anticipatory grief? Will it make grieving for her father more difficult or less difficult?

6. ______________________________

____

CARE OF THE DYING CLIENT

What state of awareness do the doctor, nurse, family and client share in this situation with Ray Morris' family?

7. ______________________________

Should Mary have answered Ray in the way that she did? Describe the role of the nurse when the client asks, "Am I going to die?"

8. ______________________________

____

June 28:
When I came on duty this morning I found Ray in severe pain. I did everything I could to make him comfortable but nothing worked. I know Ray is dying and in pain, but it hurts when he's so unkind. He called me a studpid blond and told me to get out. Maybe I should ask for a change in assignment.

Why is Ray responding in this manner? How should the nurse react when a dying client is angry and hostile?

9. ____

____

August 6:
Ray has been in such good spirits lately. Last week he told me that he wanted to see his granddaughter's first Christmas. He was so optimistic and hopeful. Today he refused to eat and slept most of the day. Is he taking a turn for the worse?

Discuss Ray's reaction. Why is he reacting in this manner?

10. ____

____

Ray is moving through the stages of dying. Describe the client's reaction to the dying process in the table below using the Kubler-Ross's model.

STAGES OF DYING	DESCRIPTION
11. Denial	
12. Anger	
13. Bargaining	
14. Depression	
15. Acceptance	

August 10:
Ray's wife and daughter visited today while I was on duty. His wife literally attacked me. She said, "Ray hasn't had a decent bath since he's been in the hospital. Don't you know how to give a bath?" How could she talk to me like this? I've given her husband excellent care.

Why did Ray's wife react this way? How should Mary have responded?

16. ____

____

August 12:
Ray's condition is deteriorating. I'm having such trouble accepting his death. Why am I feeling as bad as I did when Dad died? Will it be this way every time I care for a dying client?

Discuss Mary's reaction to her dying cient. Why is she having difficulty accepting his death?

17. ____

____

August 15:
Ray's family approached me today. They're concerned because he seems so distant and unapproachable. His daughter said, "Have we done something to offend Dad? Why does he seem so strange?"

Why is Ray relating to his family in this way? Is this a normal reaction in a dying client?

18. ____

____

August 20:
Ray seem unusually depressed today. I sat down by his bedside and said, "You seem troubled." He answered, "My wife and daughter need me so much. How can I leave them? They won't deal with the fact that I'm dying!"

What would you say to Ray? How could you help Ray's family cope with his impending death?

19. ____

____

August 25:
Ray says he wants to die at home. He's insistent on leaving the hospital. I can't support his wishes when I know his family will not admit he's dying. Maybe the local Hospice Association can help.

Discuss the concept of hospice care. How could a hospice help Ray and his family?

20. ____

____

August 28:
Ray is going home. A team member from the hospice program spent time with Ray, his doctor, and family. When the conference was over I sense a relief in Ray and his wife. They've come a long way in dealing with death.......and so have I!

Develop two nursing diagnoses for the client who is dying. Then develop at least one outcome criteria and one nursing intervention for each diagnoses.

NURSING DIAGNOSIS	CLIENT GOAL/ OUTCOME CRITERIA	NURSING INTERVENTION AND RATIONALE
21.		
22.		

CARE OF THE BODY AFTER DEATH

Complete the following statements about the care of the body after death.

23. As soon as possible after the death of the client the nurse should prepare the body. This includes ____

____

____

____

SELF ASSESSMENT QUESTIONS

24. Mary still grieves for her father six months after his death. How long will it be before Mary can reintegrate and stabilize her feelings? Approximately

    A. 9 to 24 months.
    B. 24 to 36 months.
    C. 36 to 48 months.
    D. five years or more.

25. Which of the following statements indicates that Mary has begun to recover from her father's death?

    A. "I miss Dad, but life must go on."
    B. "I still visit his grave every day."
    C. "I choke and cry every time I think about him."
    D. "I can't talk about him. It's too soon."

26. When a person grieves before the loss actually occurs, the loss is categorized as

    A. actual.
    B. perceived.
    C. expected.
    D. anticipatory.

27. When a client dies, the nurse's intial nursing action should be to

    A. clean and position the body in a natural and comfortable manner.
    B. ask family for permission to do an autopsy or donate organs.
    C. allow the family to view the body.
    D. call the clergy and notify the mortician.

28. When a client is near death the nurse should

    A. administer analgesics subcutaneously to relieve pain.
    B. administer alcoholic beverages to stimulate appetite.
    C. keep room lights very dim to prevent eyestrain.
    D. speak softly or whisper to prevent disturbing the client.

ADDITIONAL LEARNING ACTIVITIES

1. Interview a nurse who has assisted dying clients and their families. What were identified as the client's and family's needs? How did the nurse feel about providing this care?

2. Interview clients who have encountered some loss, such as a loss of mobility due to a fractured hip. How did the client perceive this loss? What were the client's greatest needs?

3. In a group of your peers, discuss your feelings about caring for a dying client.

35 MOBILITY AND IMMOBILITY

From the first few weeks in the womb until an individual is stilled by injury, disease or death, the body is in constant, dynamic motion. When a client's capacity for movement is taken away, his or her body may become ravaged by a variety of complex, interrelated complications that could result in permanent disability, chronic pain, and even death. The extent to which these complications affect a client largely depends on the knowledge and skills that the nurse brings into play while the client's ability to move is impaired. This chapter focuses on the knowledge and skills that the nurse must possess to care for clients who have impaired or limited movement. After completing this chapter the student will be able to:

- Describe the concepts of mobility and immobility.
- Identify factors that affect a person's mobility.
- Identify physiologic responses to immobility.
- Describe psychosocial responses to immobility
- Describe the etiology and pathogenesis of pressure sores.
- Identify essential data required to assess a client's mobility status.
- Identify clients at risk of developing pressure sores.
- Develop nursing diagnoses related to the client's mobility problems.
- Develop goals and outcome criteria for specific diagnoses.
- Plan and implement nursing interventions that prevent the problems of immobility.

PHYSICAL MOBILITY AND IMMOBILITY

List the factors that affect mobility.

1. ______________________________

2. ______________________________

3. ______________________________

4. ______________________________

PHYSIOLOGIC RESPONSES TO IMMOBILITY

Match the terms in column I with the appropriate description in Column II.

5. ______ Disuse osteoporosis
6. ______ Fibrosis
7. ______ Ankylosis
8. ______ Valsalva maneuver
9. ______ Orthostatic hypotension
10. ______ Virchow's triad
11. ______ Embolus
12. ______ Hypostatic pneumonia
13. ______ Negative nitrogen balance
14. ______ Natriuresis
15. ______ Renal calculi
16. ______ Urinary incontinence

F. Calcium extraction from bones resulting in decreased bone mass
J. An increase in the amount of fibrous connective tissue
B. Occurs whenever joints are not moved normally
L. Increases intrathoracic pressure.
A. May occur when an immobilized person attempts to stand
E. Factors that predispose a client to thrombophlebitis
C. A thrombus that breaks loose from the vein wall and enters circulation
D. Caused by static respiratory secretions
I. An imbalance between anabolism and catabolism
F. Increased excretion of sodium in the urine
H. Stones that develop in the kidney pelvis
K. Involuntary urination

ETIOLOGY AND PATHOGENISIS OF PRESSURE SORES

Describe the three causes of pressure sores.

17. ____

18. ____

19. ____

ASSESSING MOBILITY AND IMMOBILITY

Identify the range-of -motion movements depicted on the next page.

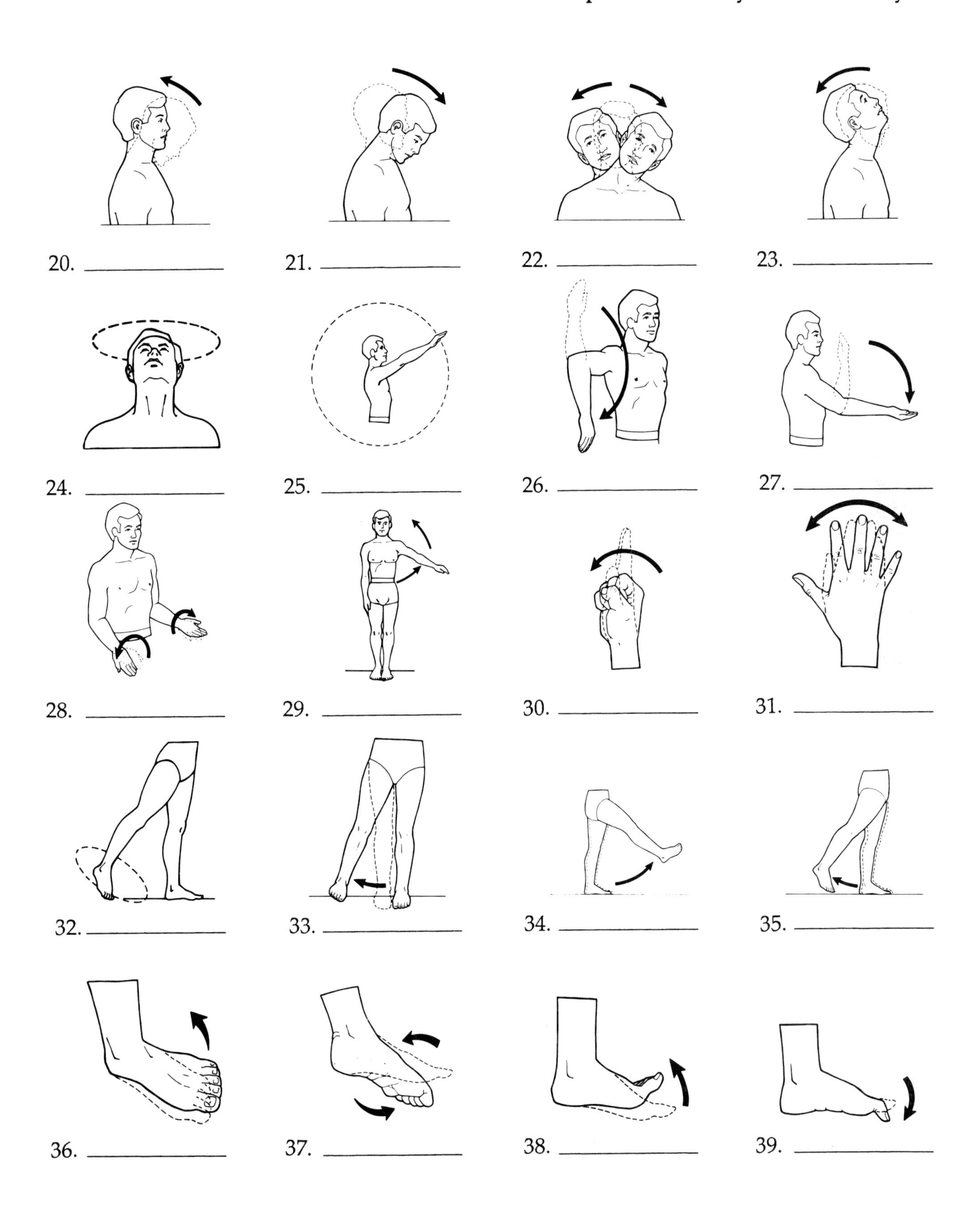
20. ____________
21. ____________
22. ____________
23. ____________
24. ____________
25. ____________
26. ____________
27. ____________
28. ____________
29. ____________
30. ____________
31. ____________
32. ____________
33. ____________
34. ____________
35. ____________
36. ____________
37. ____________
38. ____________
39. ____________

In the table below identify four categories of clients who are at risk for problems of immobility and four catergories of clients who are at risk for pressure areas.

IMMOBILITY PROBLEMS	PRESSURE AREA PROBLEMS
40.	44.
41.	45.
42.	46.
43.	47.

USING THE NURSING PROCESS

Develop two nursing diagnoses for the client who is immobilized. Then develop at least one outcome criteria and one nursing intervention for each diagnoses.

NURSING DIAGNOSIS	CLIENT GOAL/ OUTCOME CRITERIA	NURSING INTERVENTION AND RATIONALE
48.		
49.		

SELF ASSESSMENT QUESTIONS

50. A primary disability is one that is

    A. a direct result of disease or trauma.
    B. an indirect result of disease or trauma.
    C. a complication of disuse.
    D. a complication of immobility.

51. Virchow's triad describes the factors that predispose to

    A. renal calculi.
    B. hypotension.
    C. decubitus ulcers.
    D. thrombophlebitis.

52. When a client is immobilized, changes in mental function may include

    A. amnesia.
    B. anorexia.
    C. insomnia.
    D. analgesia.

53. The diet for the immobilized client should be high in

    A. protein, calories, and fiber.
    B. fat, vitamins, and calories.
    C. carbohydrates, fat, and fiber.
    D. protein, fat, and carbohydrates.

54. The urine can be acidified by including foods in the diet that lower the urine pH such as

    A. milk.
    B. cola.
    C. cranberry juice.
    D. apple juice.

ADDITIONAL LEARNING ACTIVITIES

1. Observe normal posture and motor activity in the following groups: newborns or infants, toddlers, children, adolescents, adults, pregnant women, and older adults. Note the similarities and differences.

2. Using a student as client, practice putting the client through the complete range-of-motion exercises.

3. Assess your own physical fitness in the areas of joint flexibility, muscle tone, and endurance.

36 ACTIVITY / EXERCISE

Chapter 36 focuses on the reciprocal relationship between activity, exercise and body mechanics in the maintenance of musculoskeletal health and the prevention of injury for both the nurse and the client. After completing this chapter the student will be able to:

- Identify the importance for both clients and nurses of using good body mechanics.
- Describe the importance of good body alignment for clients and nurses.
- Describe how musculoskeletal function and voluntary and involuntary muscle and reflex activity affect movement.
- Identify factors that influence body mechanics, ambulation, and alignment.
- Identify occupational groups at risk for back injury.
- Describe ways to prevent back injury.
- Identify structural abnormalities that affect body mechanics and ambulation.
- Identify ways to determine the client's capabilities and limitations for movement.
- Identify criteria used to assess a client's gait.
- Describe assessment criteria for the alignment of adults in standing, sitting, and various bed-lying positions.
- State nursing diagnoses for clients with alignment and ambulation problems.
- State outcome criteria for evaluating client responses to nursing interventions.
- Describe nursing interventions to maintain, promote, or restore normal body mechanics, alignment, and ambulation.
- Describe how to move and turn a client in bed and to transfer a client from a bed to a chair or stretcher.

BODY MECHANICS

Complete the following sentences.

1. Good body mechanics are ____

   ____

2. Body mechanics are important for clients and nurses because ________________

   ____

3. Body mechanics involves three basic elements that are ______________________

____

4. Balance is ____

5. The line of gravity is ____

6. The center of gravity is ____

List and briefly describe the six righting reflexes that keep humans upright.

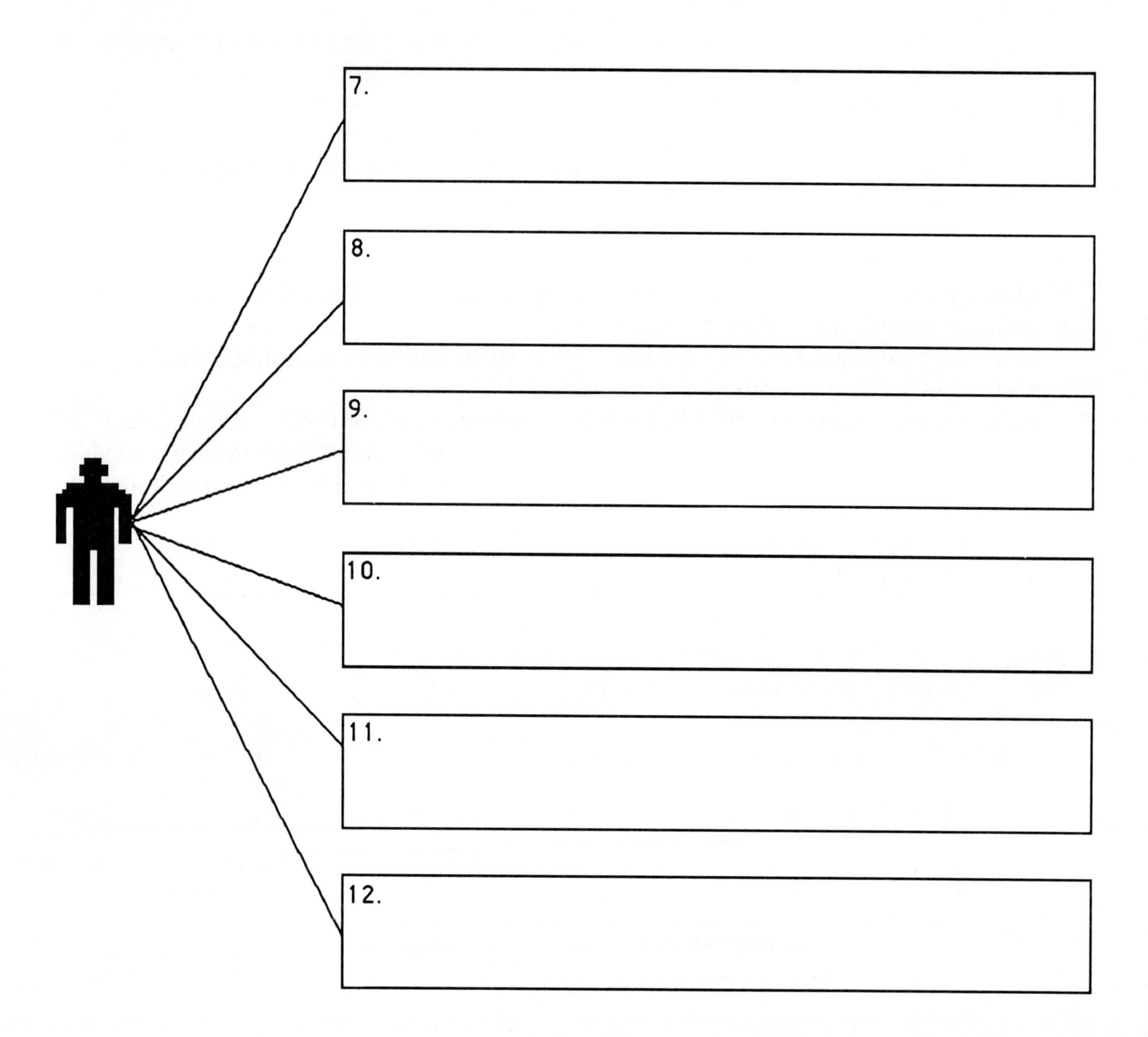

PRINCIPLES OF BODY MECHANICS

Draw a line from the term in column I to the appropriate definition in column II.

I	II
13. Lever	A. The energy or power required to accomplish movement
14. Force	B. The tendency of an object at rest to remain at rest and an object in motion to remain in motion
15. Friction	C. Force that opposes the motion of an object as it is slid across the surface of another object
16. Inertia	D. A rigid piece that transmits motion or force.
17. Fulcrum	E. A fixed point about which a lever moves

List the twelve principles related to body mechanics.

18. ______________________________

____

19. ______________________________

____

20. ______________________________

____

21. ______________________________

____

22. ______________________________

____

23. ______________________________

____

24. ____

____

25. ____

____

26. ____

____

27. ____

____

28. ____

____

29. ____

____

Identify the structural abnormalities depicted in the diagrams below.

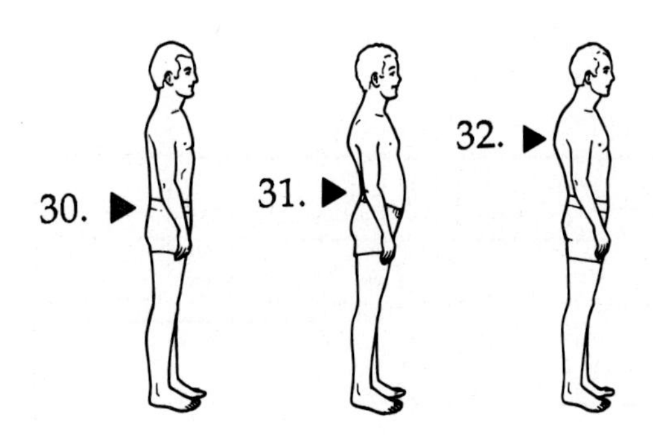

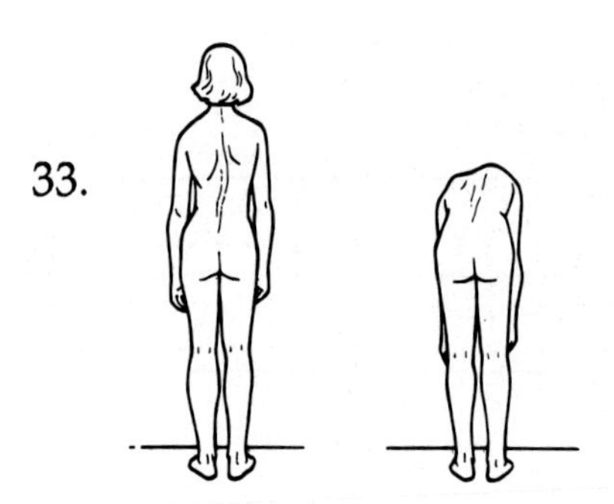

30. ____

31. ____

32. ____

33. ____

FACTORS THAT INFLUENCE BODY MECHANICS

Define each of the following terms and describe how each one contributes to poor body mechanics.

34. Osteoporosis ____

____

35. Osteoarthritis ____

____

36. Muscle atrophy and joint stiffness ____

____

USING THE NURSING PROCESS

Develop two nursing diagnoses for the client who has a mobility or immobility problem. Then develop at least one outcome criteria and one nursing intervention for each diagnoses.

NURSING DIAGNOSIS	CLIENT GOAL/ OUTCOME CRITERIA	NURSING INTERVENTION AND RATIONALE
37.		
38.		

SELF ASSESSMENT QUESTIONS

39. The center of gravity is

    A. the state of equilibrium resulting in good body alignment.
    B. an imaginary vertical line drawn through the center of a person's body.
    C. the foundation on which an object rests.
    D. the point at which all of the mass of an object occurs.

40. When assessing for scoliosis, look for

    A. an exaggerated curvature of the lumbar spine.
    B. a flexion deformity of the thoracic spine.
    C. a lateral curvature of the spine.
    D. hyperextension of the cervical spine.

41. When placing the client in high Fowler's position the nurse should

    A. flex the client's knees with the knee gatch of the hospital bed.
    B. place a small pillow under the lumbar region of the back.
    C. position the legs in external rotation.
    D. elevate the head of the bed 5 ° to 10°.

42. When logrolling a client the nurse should

    A. move the upper part of the body first, then the lower part.
    B. move the lower part of the body first, then the upper part.
    C. move the body in unison.
    D. do none of the above.

43. When transferring a client between a bed and a wheelchair the nurse should

    A. lower the bed to its lowest position.
    B. place the wheelchair at a 90° angle to the bed.
    C. stand behind the client.
    D. lift the client into the chair.

ADDITIONAL LEARNING ACTIVITIES

1. In the laboratory, practice placing a client in various bed lying positions, log rolling, and tranferring him or her between the bed, and wheelchair, bed and stretcher.

2. Develop a teaching plan for a client who needs to learn how to use a walker, crutches, or a cane.

3. Practice lifting and moving techniques. Ask a classmate to critique your techniques pinpointing breaks in body mechanic technique. If possible have someone videotape you lifting and moving a client so that you can observe how well you use body mechanics.

37 REST AND SLEEP

An important component of any client's plan of care is the consideration for rest and sleep. Nurses assess the client's capacity for rest and sleep by observing physical changes, taking a sleep history, and gathering data from a sleep diary kept by the client. In addition, the nurse should consider the client's age and stage of development, the effects of illness or injury, the amount of stimuli present in the sleeping environment, family and work responsibilities, and the client's level of stress and anxiety. Nursing interventions designed to enhance sleep include decreasing environmental stimuli, promoting relaxation, and maintaining regular bedtime routines. This chapter focuses on the knowledge and skills the nurse needs to foster rest and sleep in the client. After completing this chapter the student will be able to:

- Explain the physiologic basis of sleep.
- Identify the characteristics of NREM and REM sleep.
- Identify the four stages of NREM sleep.
- Identify the developmental variations in sleep patterns
- Identify interventions that promote sleep at various ages.
- Identify factors that affect normal sleep.
- Define terms related to common sleep disorders.
- Identify the components of a sleep assessment.
- Identify implementations that promote normal sleep.
- Describe outcome criteria for evaluating a client's response to interventions employed to promote sleep.

REST AND SLEEP

Define or describe the following terms related to rest and sleep.

1. Rest ____

____

2. Sleep ____

____

3. Biorhythms ____

____

PHYSIOLOGY OF SLEEP

True or False

4. ______ The passive theory of sleep proposes some sort of center that causes sleep by inhibiting other parts of the brain.

5. ______ The RAS is associated with the body's state of alertness and receives sensory input.

6. ______ The BSR's activity decreases with sleep.

7. ______ The circadian rhythm repeats every 24 hours.

8. ______ The infradian rhythm repeats in minutes or hours.

9. ______ NREM sleep is the same as slow-wave sleep.

10. ______ The four stages of NREM sleep last approximately two hours.

11. ______ REM sleep is thought to restore a person mentally.

12. ______ REM means rapid eye movements.

13. ______ Circadian synchronization means that a person's biological clock and sleep-wake patterns coincide.

Describe the four stages of NREM sleep in the space provided.

14. Stage I

15. Stage II

16. Stage III

17. Stage IV

SLEEP VARIATIONS ACCORDING TO AGE

Indicate the normal number of hours of sleep and the percentage of REM sleep for each developmental level in the table below.

DEVELOPMENTAL LEVEL	NUMBER OF HOURS OF SLEEP	PERCENTAGE OF REM SLEEP
18. Newborn		
19. Toddler		
20. Preschooler		
21. School-age		
22. Adolescent		
23. Young adult		
24. Middle-age		
25. Elderly adult		

FACTORS AFFECTING NORMAL SLEEP

List seven factors that affect the quality and quantity of sleep.

26. ______________________________

27. ______________________________

28. ______________________________

29. ______________________________

30. ______________________________

31. ______________________________

32. ______________________________

COMMON SLEEP DISORDERS

Match the terms in column I with the appropriate description in column II.

33. ______ Insomnia
34. ______ Hypersomnia
35. ______ Narcolepsy
36. ______ Cataplexy
37. ______ Sleep apnea
38. ______ Somnambulism
39. ______ Night terrors
40. ______ Nocturnal enuresis
41. ______ Bruxism

A. The periodic cessation of breathing during sleep

B. A sudden wave of overwhelming sleepiness that occurs during the day

C. Bedwetting

D. Sleepwalking

E. The inability to obtain an adequate amount or quality of sleep

F. Partial or complete muscle paralysis

G. Excessive daytime sleep

H. Teeth grinding during sleep

I. Horrifying dreams

SLEEP DEPRIVATION

Describe the causes and signs of sleep deprivation in the table below.

TYPE	CAUSES	SIGNS AND SYMPTOMS
42. REM deprivation		
43. NREM deprivation		

USING THE NURSING PROCESS

Describe three types of sleep data gathered during sleep history, sleep diary and a physical examination.

44. Sleep history	
45. Sleep diary	
46. Physical exam	

Develop two nursing diagnoses for the client who has problems sleeping. Then develop at least one outcome criteria and one nursing intervention for each diagnoses.

NURSING DIAGNOSIS	CLIENT GOAL/ OUTCOME CRITERIA	NURSING INTERVENTION AND RATIONALE
47.		
48.		

SELF ASSESSMENT QUESTIONS

49. A quiet nonstimulating environment promotes sleep because it

    A. allows social withdrawal.
    B. decreases cerebral arousal
    C. increases serotonin levels.
    D. elevates body temperature.

50. Administration of a backrub is a common bedtime routine because it

    A. decreases respirations.
    B. prevents dry skin.
    C. Promotes comfort.
    D. increases REM sleep.

51. The best nursing intervention to promote sleep for a client scheduled for a complicated diagnostic test is

    A. a visit from a clergyman.
    B. a careful explanation of the test.
    C. a cup of coffee or tea.
    D. the administration of a sedative.

52. Narcolepsy is

    A. a sudden attack of daytime sleepiness.
    B. difficulty breathing during sleep.
    C. a sleep disorder that includes a series of nightmares.
    D. a condition in which the adult client sleeps 16 hours per day.

53. The physiologic changes that occur during sleep include

    A. decreased gastrointestinal action.
    B. dedreased arterial pressure.
    C. relaxed venous tension.
    D. decreased basal metabolism rate.

ADDITIONAL LEARNING ACTIVITIES

1. Interview a school-age child, adolescent, adult, and elderly person. Collect the following information and compare the results:
   A. How much sleep does the person need?
   B. When does the person go to bed?
   C. What rituals, if any, help the person sleep?

D. Does the person ever have any difficulty sleeping? If so, what helps the person get to sleep.

2. Keep a sleep diary for yourself for one week. Answer the questions in item I for yourself. After analyzing your diary, determine what factors interfere with your own rest and sleep.

3. Visit a sleep laboratory at a local hospital or university. Ask to observe tracings of client who have various sleep disorders.

38 COMFORT AND PAIN

Nursing evolved out of the need to help those who were physically suffering. As a result, professional nurses have always had the prime responsibility for relieving the client's pain. When caring for the client in pain, the nurse has the opportunity to assess related symptoms, decide on appropriate nursing interventions, and evaluate the effectiveness of pain-relieving measures. This chapter focuses on the knowledge and skills that the nurse needs to care for the client experiencing pain. After completing this chapter the student will be able to:

- Identify various types of pain.
- Describe pain pathways to the brain.
- Identify physiologic manifestations of the response to pain.
- Describe subjective and objective data to be collected and analyzed when assessing pain.
- Identify examples of possible nursing diagnoses for clients with pain.
- List ways of decreasing factors that amplify the pain experience.
- Explain methods used to reduce pain intensity.
- Identify situations in which relaxation techniques can relieve pain effectively.
- Describe selected skin stimulation techniques used to relieve pain.
- Identify methods used to control intractable pain.
- Describe selected medical interventions to control pain.
- State outcome criteria by which to evaluated a client's response to interventions for pain.

TYPES OF PAIN

Describe the following types of pain.

1. Acute pain ____

   ____

2. Chronic pain ____

   ____

3. Intractable pain ____________________

____

4. Phantom pain ____________________

____

5. Radiating pain ____________________

____

6. Somatic pain ____________________

____

7. Visceral pain ____________________

____

8. Referred pain ____________________

____

PAIN EXPERIENCE

Complete the following statements about the pain experience.

9. The three stages of the pain experience are the____________________

____

10. The pain threshold is ____________________

11. Another term for pain threshold is ____________________

12. The endogenous opiods are ____________________

13. Pain tolerance is ____________________

14. Pain reaction is ____________________

15. The universal pain stimulus is ____________________

PAIN THEORIES

Briefly describe four theories that propose how pain is transmitted and perceived.

16.	17.
18.	19.

USING THE NURSING PROCESS

Develop two nursing diagnoses for the client who is experiencing pain. Then develop at least one outcome criteria and one nursing intervention for each diagnosis.

NURSING DIAGNOSIS	CLIENT GOAL/ OUTCOME CRITERIA	NURSING INTERVENTION AND RATIONALE
20.		
21.		

SELF ASSESSMENT QUESTIONS

22. Acute pain may be defined as

    A. long lasting with a slow onset.
    B. short acting with an abrupt onset.
    C. pain felt in a body part that is not present.
    D. pain that is not relieved by usual methods.

23. Pain threshold is

    A. the amount of stimuli required before a person feels pain.
    B. the maximum amount of pain a person can endure.
    C. excessive sensitivity to any amount of pain.
    D. regulated by the amount of serotonin in the body.

24. Endogenous opioids

    A. regulate the speed of C pain fibers.
    B. can be found along the spinothalamic tract.
    C. are body chemicals that modify pain.
    D. create cutaneous stimulation at nerve endings.

25. Analgesic ointments are examples of

    A. guided imagery.
    B. relaxation techniques.
    C. distraction.
    D. cutaneous stimulation.

26. Continuous morphine infusion is used for clients

    A. on the first day of surgery.
    B. who have suffered a head injury.
    C. near the end of a terminal illness with severe pain.
    D. who are afraid of receiving injections.

ADDITIONAL LEARNING ACTIVITIES

1. Talk to a client who has had pain recently. How was the pain relieved? Which nursing activities helped relieve the pain and which did not? How did the client describe the pain?

39 NUTRITION

Whether the nurse is performing a general physical assessment or specifically searching for nutritional status data, the client must be carefully scrutinized for overt and covert signs and symptoms of malnutrition or other nutritional disorders. Once the client's problems have been assessed and diagnosed, the nurse collaborates with the diet therapist, physician, and client to promote, maintain, and restore optimum nurtritional status. This chapter focuses on the knowledge and collaborative skills that the nurse must possess while applying the nursing process to clients with alterations in nutritional status. After completing this chapter, the student will be able to:

- Describe nutrition, metabolism, and energy requirement.
- Identify functions and food sources of selected nutrients and some clinical signs of deficiency and excess.
- Describe the use of daily food group guides.
- Identify potential nutritional problems of vegetarians and suggest ways to avoid them.
- Describe necessary dietary modifications for older adults.
- Identify clinical signs of inadequate nutritional status.
- Describe a format for nutritional assessment.
- Identify factors that influence a person's eating patterns.
- Identify nursing diagnoses and factors contributing to the client's nutritional status.
- Identify interventions to stimulate a client's appetite.
- Describe ways to assist clients with meals.
- Describe some aids that enable self-feeding.
- Discuss some special nutritional services available for selected subgroups of the population.
- Recognize characteristics of commonly prescribed diets.
- Identify nursing responsibilities in administering enteral and parental nutriton.

NUTRITION AND METABOISM

Complete the following sentences.

1. ____________________ is the sum of all interactions between an organism and the food it consumes.

2. ____________________ are the organic and inorganic chemicals found in foods and required for proper body functioning.

3. ____________________ is the nutrient content of a specified amount of food.

4. ____________________ is a unit of heat energy.

5. ____________________ is the amount of heat required to raise the temperature of 1 kg of water 1 degree C and is the unit used in nutrition.

6. ____________________ the rate at which the body metabolizes food to maintain the energy requirements of a person who is awake and at rest.

ESSENTIAL NUTRIENTS

Fill in the blank concerning the essential nutrients.

7. Carbohydrates are composed of three elements: ______________________________

8. Carbohydrates consist of two basic types of sugars which are called:____________

9. Carbohydrates that come from the earth are categorized as ____________________

10. Carbohydrates that are extracted from their natural sources and added to food are called

    ____

11. A carbohydrate that is derived from plants and cannot be digested by humans is called ____________________

12. The biological catalysts that speed up chemical reactions are ____________________

13. Proteins are substances that upon digestion yield ____________________________

14. Proteins that cannot be manufactured by the body are ________________________

15. Complete protein foods are ____

16. When nitrogen input exceeds output, the result is____________________________

17. Fats can be categorized as ____

18. Hyperlipidemia puts people at risk for ______________________________

19. Vitamins A, D, E and D are categorized as ______________________________

20. Eighty percent of the mineral content of the body is ______________________

21. Iron, zinc, and iodine are called ______________________________

FACTORS INFLUENCING DIET

Case Study:

Gertrude Morris was tired! - so tired she barely had the strength to get out of bed and start the day. She couldn't understand why she felt this way. . . . perhaps she needed something to eat. She'd make herself a nice breakfast, read the paper, and began the day right. but when Gertrude opened the refrigerator, she found a rancid package of bacon and half of a coffee cake that had gone stale. "That's alright," she thought, "I'll have a bowl of cereal and some fruit," but then she remembered there was no milk. "I've got to get out to the grocery store," she mused, "but I just hate to take that smelly bus downtown, I think I'll just wait until later and order a pizza. I'll drink some coffee and read the paper," but the coffee tasted strange and the newsprint blurred. "I've got to get my eyes checked soon," she thought. Before Gertrude had time to order the pizza that afternoon, she became weak and fell down the basement steeps. When her neighbor found her lying dazed and bruised at the bottom of the steps the next morning, Gertrude was brought to the hospital. The nurse admitting Gertrude made the following notations: 76 year-old frail, confused, white, dehydrated, pale, edentulous female with reduced turgor, admitted for observation because of fall in home: height 153 cm; weight 48 kg; blood pressure 98/40; pulse 140; respirations 36.

How would you rate Gertrude's nutritional status? Describe the clinical signs and symptoms that support your evaluation.

22. ____

____

Discuss the underlying physiologic causes for malnutrition in elderly clients like Gertrude.

23. ____

____

Discuss Gertrude's poor nutritional status from psychosocial and economic viewpoints.

24. ____

____

Gertrude tells you she likes pasta, Chinese food, and beef. She says she hates pork and fish. "I'll eat fresh vegetables and fruits if I can get them," she says, "but I have a difficult time getting to the store." Will her nutritional problems be solved with this diet? If your answer is yes, provide a rationale. If your answer is no, explain why.

25. ____

____

USING THE NURSING PROCESS

Develop two nursing diagnoses for the client who is immobilized. Then develop at least one outcome criteria and one nursing intervention for each diagnosis.

NURSING DIAGNOSIS	CLIENT GOAL/ OUTCOME CRITERIA	NURSING INTERVENTION AND RATIONALE
26.		
27.		

SELF ASSESSMENT QUESTIONS

28. Gertrude needs more calcium and vitamin C in her diet. You should encourage her to eat more

    A. green vegetables and milk.
    B. liver, meats and whole grains.
    C. milk, wheat germ, and liver.
    D. fruits, milk, and dairy products.

29. The quality of Gertrude's diet would improve if she would:

    A. increase vegetables and decrease milk.
    B. increase sugar and decrease fat.
    C. increase fat and decrease protein.
    D. increase grains and decrease salt.

30. Elderly people like Gertrude should:

    A. maintain protein intake.
    B. reduce mineral intake.
    C. increase carbohydrate intake.
    D. decrease fiber intake.

31. Gertrude says she is planning to go on a vegetarian diet and asks your advice. Which of the following reaponses would be most helpful to Gertrude. "A vegetarian diet

    A. would be a serious affront to your gastrointestinal system at your age."
    B. has serious deficiencies in proteins because of the lack of meat, poultry, and fish."
    C. will be healthy if you will eat twice the amount of food you eat now to meet your daily caloric requirements."
    D. can be healthy as long as you include legumes, grains, fruits, vegetables, and milk products in your diet."

32. Gertrude's decubitus ulcer could be a result of deficiency in vitamin:

    A. A
    B. B
    C. C
    D. D

ADDITIONAL LEARNING ACTIVITIES

1. List your own dietary intake over a 3-day period. Analyze it for caloric and nutrient value. Determine whether your diet meets federal guidelines.

2. Interview the parents of a toddler, school-age child, and adolescent about their child's eating habits. How do the eating habits at each age differ?

3. Interview a member of a religious group that advocates a vegetarian diet. Investigate how he or she prepares their food. What special food preparation is involved? Do members experience any health problems?

40 FLUIDS AND ELECTROLYTES

An important aspect of caring for a client who is ill, is to integrate the concepts of homeostasis with a thorough knowledge of fluid and electrolyte balance. The nurse accomplishes this by assessing and diagnosing the client's response to changes in fluid and electrolyte levels, as well as planning outcome criteria that are aimed at correcting any imbalances that may occur. Independent and dependent nursing interventions are implemented that concentrate on promoting, maintaining, or restoring normal fluid and electrolyte balance. This chapter focuses on the knowledge and skill that the nurse must possess to apply the nursing process to the care of clients with alterations in fluid and electrolyte balance. After completing this chapter the student will be able to:

- Describe factors affecting the proportion of the body weight that is fluid.
- Identify the major electrolytes of the intracellular and extracellular fluid compartments and body secretions.
- Describe how fluids and electrolytes move through the body.
- Explain how the osmotic and hydrostatic pressures influence movement of fluid through membranes.
- List factors that influence fluid and electrolyte balance.
- Describe how body mechanisms regulate fluid and electrolyte balance.
- Describe the role of the lungs and kidneys in regulating acid-base balance.
- Describe the significance of diagnostic tests used to monitor fluid, electrolyte, and acid-base imbalances.
- Recognize clinical signs and laboratory findings of selected fluid and electrolyte imbalance.
- Describe four primary acid-base disturbances.
- List examples of nursing diagnoses related to fluid, electrolyte, and acid-base balance.
- State outcome criteria for evaluating the client's responses to strategies implemented to promote fluid and electrolyte balance.
- Assist clients to modify their fluid intake.
- Monitor and regulate intravenous infusions.
- Describe how to change intravenous containers and tubing.
- Give guidelines for discontinuing an intravenous infusion.
- Identify potential problems and risks and blood transfusions.
- Explain basic purposes of and sites used for total parenteral nutrition.

DISTRIBUTION OF BODY FLUID

Define the following terms.

1. Intracellular fluid ____

   ____

2. Extracellular fluid ____

   ____

3. Intravascular plasma ____

   ____

4. Interstitial fluid ____

   ____

5. Secretion ____

   ____

6. Excretion ____

   ____

PROPORTIONS OF BODY FLUID

Circle the correct choice in each of the following statements.

7. Fluid constitutes about (10% - 30% - 57 %) of the average healthy adult man's weight, which amounts to approximately (20 - 30 - 40) liters.

8. Infants have the (highest - lowest) proportion of fluid as compared to other age groups.

9. As people grow older, the proportion of fluid (increases - decreases).

10. In the obese individual, the less body fat present, the (greater - lesser) the proportion of body fluid.

MOVEMENT OF BODY FLUIDS AND ELECTROLYTES

True and False

11. _______ An ion is a charged particle called an electrolyte.

12. _______ Electrolytes are capable of conducting electricity.

13. _______ Ions that carry a positive charge are called anions.

14. _______ Ions that carry a negative charge are called anions.

15. _______ Diffusion is the movement of water across cell membranes.

16. _______ Isotonic means a solution has the same concentration as blood plasma.

17. _______ Hypertonic solutions have a greater concentration of solutes than plasma.

18. _______ Active transport means that substances can move across cell membranes from a less concentrated solution to a more concentrated one.

19. _______ Hydrostatic pressure is the pressure exerted by a fluid within a closed system.

REGULATING FLUID VOLUME

List the four ways body fluids are lost.

20. ____

21. ____

22. ____

23. ____

Describe the nursing interventions used when the electrolytes in the table below are in excess or in deficit. Indicate the normal serum level of the electrolyte in the electrolyte column.

ELECTROLYTE	INTERVENTIONS DEFICITS	INTERVENTIONS EXCESS
24. Sodium ______mEq/L		
25. Potassium ______mEq/L		
26. Calcium ______mEq/L		
27. Chloride ______mEq/L		
28. Magnesium ______mEq/L		
29. Phosphate ______mEq/L		

ACID-BASE BALANCE

Briefly describe how the kidneys participate in maintaining acid-base balance.

30. ____

____

USING THE NURSING PROCESS

Develop two nursing diagnoses for the client who has a fluid and electrolyte imbalance. Then develop at least one outcome criteria and one nursing intervention for each diagnoses.

NURSING DIAGNOSIS	CLIENT GOAL/ OUTCOME CRITERIA	NURSING INTERVENTION AND RATIONALE
31.		
32.		

SELF ASSESSMENT QUESTIONS

33. When a client develops symptoms of muscle weakness, speech changes, rapid or weak pulse, or abdominal distention, you should suspect

    A. hypokalemia.
    B. hyponatremia.
    C. respiratory acidosis.
    D. respiratory alkalosis.

34. If your client shows signs of a bood transfusion reaction, your immediate action should be to

    A. raise the head of the bed and start oxygen.
    B. stop the blood and start an infusion of saline.
    C. call the laboratory and have a blood sample drawn.
    D. ask the client for a urine sample.

35. An independent nursing action for the client with a fluid deficit may be

    A. intake and output.
    B. infusion of normal saline.
    C. chest x-ray.
    D. insertion of central venous catheter.

36. Signs of infiltration at an intravenous site are

    A. hematorma, increased rate of infusion, and warm skin.
    B. yellow drainage, pain, and swelling.
    C. inadequate blood return, decreased pulse, and pain.
    D. pain, swelling, and inadequate blood return.

37. An appropriate outcome criteria for the client with decreased potassium intake is that

    A. client will increase potassium intake.
    B. client will eat potassium-rich foods such as a banana at each meal.
    C. client understands the meaning of decreased potassium level.
    D. client talks about decreased potassium level frequently.

ADDITIONAL LEARNING ACTIVITIES

1. In a laboratory setting assemble and set up the equipment for starting an IV line. Check the drop rate on the administration package and regulate the flow for the following infusion orders:
   A. 1000 ml in 8 hours
   B. 2000 ml in 24 hours
   C. 100 ml per hour

2. Review the serum electrolyte laboratory test results for five clients on your clinical unit. If any of the readings are abnormal assess the clients for relevant clinical signs of deficits or excesses.

41 OXYGENATION

In the practice of professional nursing today, nurses are caring for increasing numbers of clients wih respiratory problems. Measures such as cardiopulmonary resuscitation, artificial ventilation, improved oxygen delivery systems, and the use of newer medications are increasing the chances of survival for clients with alterations in oxygenation. The nurse must know precisely what to look for when performing a respiratory assessment. Knowledge about changes in respiratory function is essential, especially when complications are a possibility. Early detection of changes in the client's condition can allow time for the nurse to intervene before a respiratory crisis occurs. Clients with respiratory disease depend on the nurse to solve problems, develop adquate plans of care, and prepare them for living with chronic respiratory problems after discharge. This chapter focuses on the knowledge and skills the nurse must possess when caring for the client with an alteration in oxygenation. After completing this chapter the student will be able to:

- Explain the three phases of respiration.
- Describe the basic mechanics of breathing.
- Identify the requirements of adequate ventilation.
- Explain mechanisms regulating the respiratory process.
- Describe factors that influence the rate of diffusion of gases through the respiratory membrane.
- Explain how oxygen is transported to the tissues and how carbon dioxide is transported from the tissues.
- Identify factors influencing respiratory and circulatory functions.
- Describe common altered breathing patterns.
- List the signs of an obstructed airway.
- Identify common responses to alterations in respiratory and circulatory status.
- Describe positions that facilitate oxygenation.
- Explain the importance of hydration in promoting adequate oxygenation.
- Describe the use of selected inhalation therapy devices and practices.
- Describe various methods to administer oxygen.
- Compare administering oxygen therapy by nasal cannula and face mask.
- State outcome criteria for evaluating client responses to implemented measures to promote adequate oxygenation.

PHYSIOLOGY AND REGULATION OF RESPIRATION

Define the process of respiration.

1. ____

____

Describe the three phases of respiration in the spaces below.

PROCESS OF RESPIRATION

2.	3.	4.

Identify the terms described in the following statements.

5. ____________________ is the normal volume of air inspired and expired.

6. ____________________ is the maximum volume to which the lungs can expand.

7. ____________________ is the pressure within the lungs.

8. ____________________ is the pressure outside or around the lungs.

9. ____________________ is the lung expansibility or stretchability.

10. ____________________ is the tendency for the lungs to collapse away from the chest wall.

11. ______________________ is a lipoprotein mixture that counterbalances the surface tension of the fluid lining the alveoli.

12. ______________________ is the movement of gases from an area of greater concentration to an area of lower concentration.

13. ______________________ is an oxygen carrying red pigment.

14. ______________________ is the amount of blood pumped by the heart.

15. ______________________ are red blood cells.

16. ______________________ is the percent of the blood that are erythrocytes.

17. ______________________ is a number of groups of neurons located in the medulla oblongata and pons.

18. ______________________ is a group of cells that when stimulated cause inspiration.

FACTORS AFFECTING OXYGENATION

List the six factors that affect oxygenation.

19. ____

20. ____

21. ____

22. ____

23. ____

24. ____

ALTERATIONS IN RESPIRATORY FUNCTION

List at least two of the the early and late signs of hypoxia.

Early signs of hypoxia.

25. ______________________________

26. ______________________________

Late signs of hypoxia.

27. ______________________________

28. ______________________________

Match the altered breathing pattern in column I with its description in column II.

29. _______ Eupnea
30. _______ Tachypnea
31. _______ Anoxemia
32. _______ Bradypnea
33. _______ Hyperventilation
34. _______ Kussmaul breathing
35. _______ Hypoventilation
36. _______ Cheyne-Stokes
37. _______ Alveolar hyperventilation
38. _______ Apneustic
39. _______ Blot's
40. _______ Dyspnea
41. _______ Orthopnea

A. An excessive amount of air in the lungs.
B. Marked rhythmic waxing and waning of respirations.
C. Normal respiration.
D. Inability to breathe except in an upright position.
E. The amount of air in the alveoli exceeds the body's metabolic requirements.
F. Inadequate alveolar ventilation.
G. The body attempts to compensate by blowing off the carbon dioxide.
H. Decreased oxygen in the blood.
I. Rapid respirations.
J. Shallow breaths interrupted by apnea.
K. Abnormally slow respirations.
L. Prolonged gasping inspiration followed by a very short expiration.
M. Difficult or labored breathing.

List the three signs of an obstructed airway.

42. ____

43. ______________________________

44. ______________________________

USING THE NURSING PROCESS

Explain why adequate hydration is important in promoting adequate oxygenation.

45. ______________________________

____

Explain the value of the following devices to the client with an oxygenation problem.

46. Sustained maximal inspiration devices ______________________________

____

47. Intermittent positive pressure breathing ______________________________

____

Develop two nursing diagnoses for the client who has an oxygenation problem. Then develop at least one outcome criteria and one nursing intervention for each diagnoss.

NURSING DIAGNOSIS	CLIENT GOAL/ OUTCOME CRITERIA	NURSING INTERVENTION AND RATIONALE
48.		
49.		

SELF ASSESSMENT QUESTIONS

50. Your first response when a client says, "I can't breathe," should be to

    A. assess for patent airway.
    B. raise the head of the bed.
    C. auscultate the chest.
    D. turn on the oxygen.

51. Which of the following oxygen delivery systems would be best for a client who is hyperventilating?

    A. nonrebreather mask
    B. Venturi mask
    C. nasal cannula
    D. rebreather mask

52. Adequate humidy must be provided for a client with a tracheostomy because humidity

    A. prevents aspiration.
    B. promotes deep breathing.
    C. liquifies secretions.
    D. prevents a fluid deficit.

53. The client who is receiving oxygen at home should understand that oxygen

    A. supports combustion.
    B. is flammable.
    C. is a contaminant.
    D. is radioactive.

54. An advantage of the Venturi mask is that it:

    A. delivers an equal amount of room air mixed with oxygen.
    B. allows rebreathing of carbon dioxide if the client hyperventilates.
    C. is more comfortable to wear than the nasal cannula.
    D. delivers precise oxygen concentrations to the client.

ADDITIONAL LEARNING ACTIVITIES

1. Assess and compare the respiratory and pulse rates of an infant, toddler, school-age child, adult and an elderly person.

2. Observe a nurse or inhalation therapist teaching a client postural drainage, deep breathing, and effective coughing. What points were emphasized?

3. Practice applying a cannula and face mask in the nursing laboratory .

4. Follow an inhalation therapist on daily rounds to observe clients receiving various types of oxygenation therapy.

42 FECAL ELIMINATION

Bowel elimination is a normal body function that can be simply regulated by proper diet, fluid intake, and exercise in the healthy client. However when the client is ill, very young or aged, fecal elimination can become a major problem. As a result an important role for the nurse is to educate the client and the family about health measures that promote, maintain, or restore healthy fecal elimination. This chapter focuses on the knowledge and skills the nurse must possess to apply the the nursing process when caring for a client with an alteration in bowel elimination.

- Describe the functions of the lower intestinal tract.
- Identify factors that influence fecal elimination and patterns of defecation.
- Distinguish between normal and abnormal characteristics and constituents of feces.
- Describe methods used to assess the intestinal tract.
- Differentiate among specific common fecal elimination problems.
- Identify common causes and effects of selected fecal elimination problems.
- Identify measures that maintain normal fecal elimination patterns.
- Relate common interventions to specific fecal elimination problems.
- Give reasons for selected nursing interventions.
- Describe modifications for clients with ostomies.
- State outcome criteria essential for evlauating the client's progress.

PHYSIOLOGY OF DEFECATION

Define the following terms related to the physiology of defecation.

1. Persistalsis ____________________

   ____

2. Chyme ____________________

   ____

3. Haustral churning ____________________

   ____

In the following diagram, identify the anatomical landmarks in the spaces provided, then use a marker to trace the digestive tract from the esophagus down through the anus.

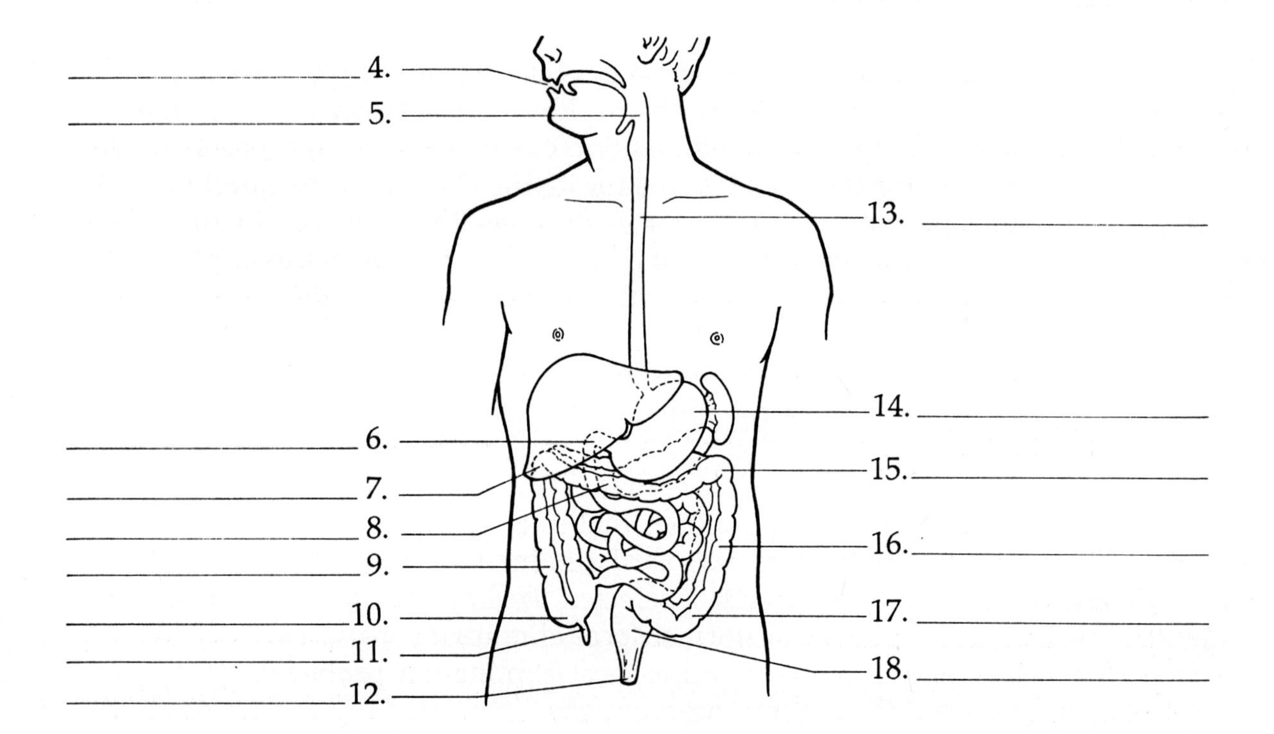

FACTORS THAT AFFECT DEFECATION

List twelve factors that affect defecation.

19. ____
20. ____
21. ____
22. ____
23 ____
24. ____
25. ____
26. ____
27. ____
28. ____
29. ____
30. ____

USING THE NURSING PROCESS

What are the six characteristics the nurse looks for when assessing feces?

31. ______________________________

32. ______________________________

33. ______________________________

34. ______________________________

35. ______________________________

36. ______________________________

Describe the causes and nursing implications for the following types of abnormal feces.

ABNORMAL FECES	POSSIBLE CAUSES	NURSING IMPLICATIONS
37. Clay-colored		
38. Black, tarry		
39. Red		
40. Green		
41. Pus		

Develop two nursing diagnoses for the client who has a problem with fecal elimination. Then develop at least one outcome criteria and one nursing intervention for each diagnosis.

NURSING DIAGNOSIS	CLIENT GOAL/ OUTCOME CRITERIA	NURSING INTERVENTION AND RATIONALE
42.		
43.		

SELF ASSESSMENT QUESTIONS

44. Which of the following should be assessed in the presence of hyperactive bowel sounds?

    A. pain, last bowel movement, and breath sounds
    B. pulse rate, and hydration
    C. abdominal masses, pain, and last bowel movement
    D. urinary elimination, pulse rate, and pain.

45. During physical examination the abdomen is auscultated prior to palpation or percussion because palpation or percussion

    A. is facilitated by auscultation.
    B. must always be done together.
    C. may temporarily produce abnormal bowel sounds.
    D. may cause abdominal trauma.

46. Elderly persons are most often prone to fecal elimination problems because

    A. they may have structural or neurological abnormalities.
    B. disease processes in the elderly always interfere with fecal elimination.
    C. their diet includes high-residue foods.
    D. they may not have a structured fecal elimination pattern.

47. If a client complains of pain while you are assessing the abdomen, you should

    A. discontinue palpation and notify the doctor immediately.
    B. assess for additional data concerning the pain.
    C. administer aspirin compound, then proceed with the examination.
    D. help the client cope with the pain.

48. If a client complains of cramping abdominal pain during the administration of an enema, the nurse should

    A. discontinue the enema immediately.
    B. clamp tubing until the cramping ceases.
    C. turn the client on the left side.
    D. administer the enema at a more rapid rate.

ADDITIONAL LEARNING ACTIVITIES

1. Note the methods used to maintain regular bowel function in elderly clients in a hospital or nursing home.

2. Select a client with a fecal elimination problem. Complete an abdominal/fecal elimination assessment for this client.

3. Select a client who has undergone a recent ostomy procedure. What kinds of problems with fecal elimination is this client experiencing? What nursing interventions are being utilized to minimize the client's problems?

43 URINARY ELIMINATION

This chapter focuses on the knowledge and skills the nurse must possess to apply the the nursing process when caring for a client with an alteration in urinary elimination.

- Describe the process of micturition.
- Identify factors that influence urinary elimination.
- Describe common alterations in urinary elimination.
- Identify common causes of selected urinary problems.
- Describe urinary diversion ostomies.
- Identify essential components of a urinary elimination nursing history.
- Describe physical examination methods used to identify elimination problems.
- Identify normal and abnormal characteristics and constituents of urine.
- Describe diagnostic measures to assess kidney function and urinary tract abnormalities.
- Explain how to collect urine specimens and perform simple tests.
- Develop nursing diagnoses related to urinary elimination.
- Develop client goals and outcome criteria for clients with nursing diagnoses related to urinary elimination.
- Describe interventions to maintain normal urinary elimination and to assist clients with urinary incontinence and retention.
- Identify ways to prevent urinary infection.
- Identify interventions required for clients with retention catheters.
- Describe interventions to maintain normal urinary elimination and assist clients with urinary retention and incontinence.

PHYSIOLOGY OF URINARY ELIMINATION

Describe the process of micturition.

1. ____

____

____

____

Write in the names of the structures of the male and female urinary tracts in the figures below.

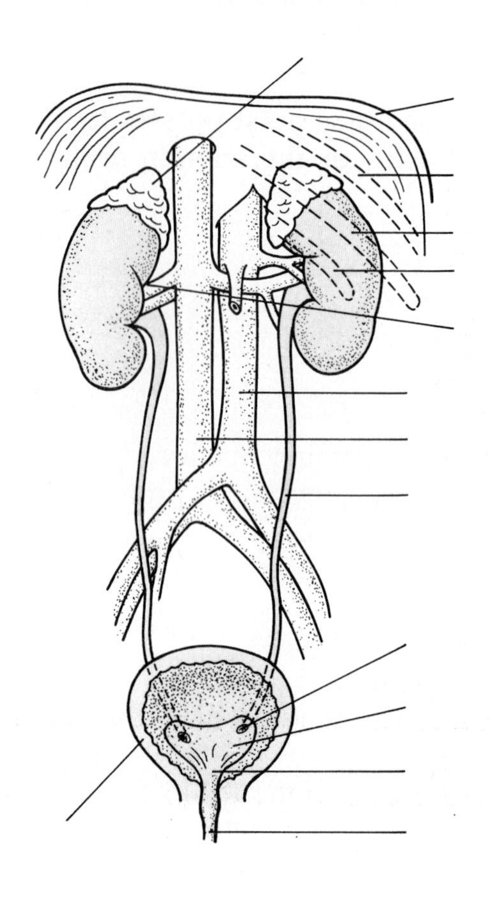

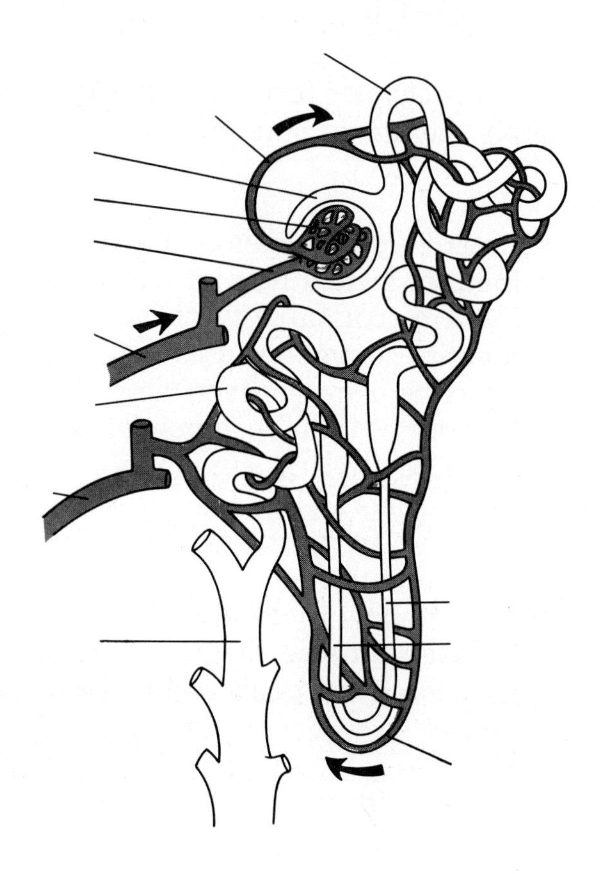

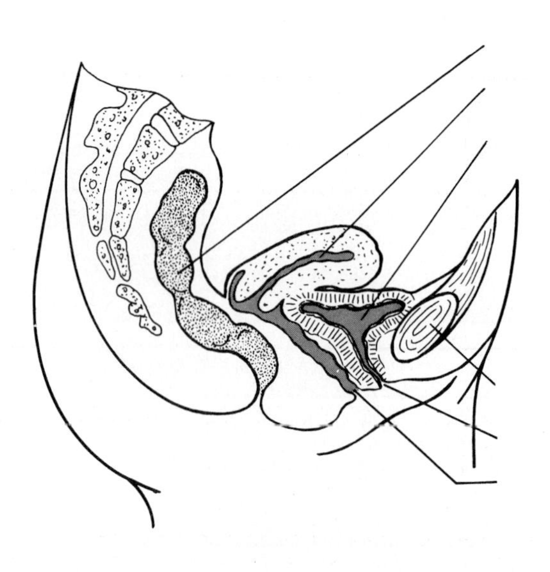

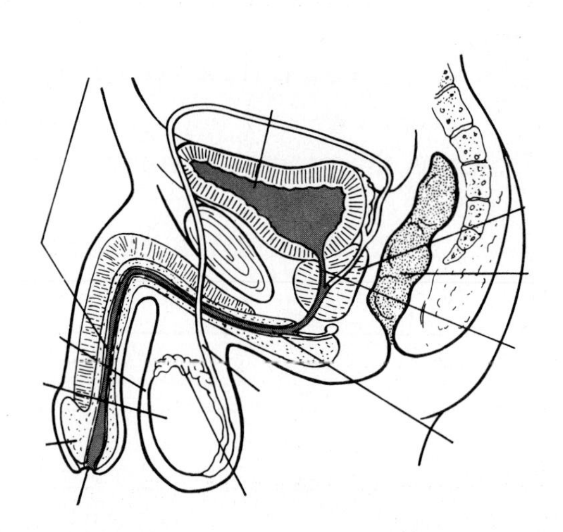

FACTORS AFFECTING VOIDING

List seven factors that affect urination.

2. ______________________
3. ______________________
4. ______________________
5. ______________________
6. ______________________
7. ______________________
8. ______________________

ALTERED URINARY ELIMINATION

Define the six common types of altered urinary elimination ,then list the causative factors.

ELIMINATION PROBLEM	DEFINITION	CAUSITIVE FACTORS
9.		
10.		
11.		
12.		
13.		
14.		

URINARY DIVERSIONS

Briefly describe the following types of urinary diversions.

15. Cutaneous ureterostomy ______________________________

16. Ileal conduit ______________________________

17. Coninent vesicostomy ______________________________

18. Ureterosigmoidoscopy ______________________________

USING THE NURSING PROCESS

What are the seven guidelines the nurse uses when assessing urinary elimination?

19. ______________________________
20. ______________________________
21. ______________________________
22. ______________________________
23. ______________________________
24. ______________________________
25. ______________________________

Indicate the normal characteristics of urine, then indicate abnormal findings and their possible causes.

NORMAL FINDINGS	ABNORMAL FINDINGS	POSSIBLE CAUSES
26. Amount		
27. Glucose		
28. pH		
28. Specific gravity		
29. Blood		

Develop two nursing diagnoses for the client who has a problem with urinary elimination. Then develop at least one outcome criteria and one nursing intervention for each diagnoss.

NURSING DIAGNOSIS	CLIENT GOAL/ OUTCOME CRITERIA	NURSING INTERVENTION AND RATIONALE
30.		
31.		

SELF ASSESSMENT QUESTIONS

32. A urinary tract infection is likely to occur when a retention catheter is in place if

    A. the catheter is disconnected from the drainage tube when getting the client out of bed.
    B. the catheter is not irrigated periodically.
    C. aseptic technique is used to insert the catheter.
    D. the drainage bag is emptied at the end of each shift.

33. Stress incontinence refers to

    A. an involuntary loss of urine associated with straining.
    B. an involuntary loss of urine associated with cystitis.
    C. an uncontrollable urge to void.
    D. accidental loss of urine due to impaired mobility.

34. An ileal coduit is more effective than a ureterostomy because

    A. the urine is diverted to the bowel.
    B. the bladder wall forms a continent pouch.
    C. There is less chance of ascending kidney infection.
    D. the ureters are reattached to the bladder at a different site.

35. Immobilized clients who are susceptible to calculi formation require

    A. 500-1000 ml per day.
    B. 1000-2000 ml fluid per day.
    C. 2000-3000 ml fluid per day.
    D. more than 3000 ml fluid per day.

36. When bladder training the incontinent adult, the nurse should

    A. reduced the daily intake of fluid to 1000 ml per day.
    B. encourage liquids such as tea or coffee.
    C. teach the client to perform Kegel's exercises frequently.
    D. encourage the client to go to the bathroom when he or she has the urge to void.

ADDITIONAL LEARNING ACTIVITIES

1. Interview a mother who has a 3 or 4-year old child. Determine how the child was, or is being bladder trained.

2. Discuss bladder retraining for an adult who has lost bladder control with a rehabilitation nurse specialist. What are the similarities and differences in bladder training an adult who has lost bladder control from that of a child who has never had control?

3. In a clinical setting develop and implement a plan of care for a client who needs bladder retraining.

4. Select a client who has undergone a recent ostomy procedure. What kinds of problems with urinary elimination is this client experiencing? What nursing interventions are being utilized to minimize the client's problems?

5. Interview the person(s) responsible for infection control in a hospital setting regarding policies for the care of clients with retention catheters.

SENSORY PERCEPTION AND COGNITION

Chapter 44 deals with the care of the client who has problems receiving or perceiving stimuli due to sensory deprivation, overload, or deficits. Special emphasis is placed on the client at risk for sensory problems as well as the clinical signs of sensory/perceptual alterations. After completing this chapter the student will be able to:

- Identify structural and physiologic elements of the sensory-perceptual process.
- Identify stages of awareness and consciousness.
- Identify causes and signs of three kinds of sensory disturbance.
- Identify factors influencing sensory function.
- Identify clients most at risk of sensory disturbances.
- Describe essential methods used to assess sensory/perceptual function and cognition.
- Identify clinical signs of sensory dysfunction.
- Develop nursing diagnoses for clients with impaired sensory and cognitive function.
- List outcome criteria to evaluate the effectiveness of goal achievement.
- Exlpain nursing interventions to maintain and promote sensory stimulation.
- Explain interventions that help the client adqpt to altered sensory function and decrease further sensory loss.

THE SENSORY PROCESS

True or False

1. ________ The sensory process involves two components: sensory reception and sensory perception.

2. ________ Sensory perception is the process of receiving stimuli or data.

3. ________ Kinesthetic refers to awareness of the position and movement of body parts.

4. ________ Gustatory is the external stimuli related to smell.

5. ________ Sensory pathways tend to travel parallel or decussate and register sensations from the opposite side of the body.

6. ________ Stereognosis is the awareness of an object's size, shape, and texture.

FACTORS AFFECTING SENSORY STIMULATION

List the six factors that affect sensory stimulation.

7. ______________________________ 10. ______________________________

8. ______________________________ 11. ______________________________

9. ______________________________ 12. ______________________________

SENSORY ALTERATIONS

Describe the following types of sensory alterations.

13. Sensory deprivation______________________________

____

14. Sensory overload ______________________________

____

15. Sensory deficit______________________________

____

USING THE NURSING PROCESS

List and briefly describe four clinical signs of sensory/perceptual alterations.

16. ______________________________

____

17. ______________________________

____

18. ____

____

19. ____

____

Develop two nursing diagnoses for the client who impaired mobility. Then develop at least one outcome criteria and one nursing intervention for each diagnosis.

NURSING DIAGNOSIS	CLIENT GOAL/ OUTCOME CRITERIA	NURSING INTERVENTION AND RATIONALE
20.		
21.		

SELF ASSESSMENT QUESTIONS

22. When talking to a hearing-impaired client the nurse should

    A. talk at a slow rate and a loud tone of voice.
    B. emphasize key words by enunciating them loudly.
    C. use short clear sentences.
    D. make sure the client can see his or her face.

23. Which of the following interventions should be used when caring for an unconscious client?

    A. change the daily schedule from day-to-day to stimulate awareness
    B. use touching and stroking as a form of communication
    C. speak in low tones or in whispers to avoid disturbing client's rest.
    D. avoid talking to the client since he or she is unable to respond.

24. When caring for a client with visual impairment the nurse should

    A. use neutral colors to prevent eye irritation.
    B. discourage handling articles that could cause trauma such as a knife.
    C. obtain reading material with large print.
    D. move furniture so that client will not fall.

25. An hallucinations are

    A. perceptions of external stimuli that are not present.
    B. misinterpretations of external stimuli.
    C. false ideas.
    D. a form of day dreaming.

26. Which of the following clients would probably experience a sensory disturbance?

    A. a ten year-old boy who is hospitalized with appendicitis.
    B. a sixteen year-old girl with a fractured ankle.
    C. a forty-five year-old man who has colitis.
    D. a sixty-seven year-old woman with bacterial pneumonia.

ADDITIONAL LEARNING ACTIVITIES

1. Experiment with sensory deprivation by wearing a blind fold or ear plugs for three hours while carrying out your normal activities. Share your experience and feelings with your classmates.

2. Visit a local hearing aid shop. Review the various types of hearing aids available to clients with hearing impairments.

3. Develop an orientation plan for a client who is confused and/or disoriented. Spend at least one hour a day for several days carrying out your plan. What behavioral changes occured as a result of your plan?

45 MEDICATIONS

The administration of medications is one of the major responsibilities of the professional nurse. The nurse who administers drugs in a knowledgeable, competent, and safe manner protects the client from harm and protects herself or himself from malpractice litigation. Safe, competent drug administration requires the nurse to possess an extensive knowledge of medications and their effects. With over 7,000 principle drugs available, the nurse cannot rely on her or his own knowledge base as a sole resource for drug information. Responsible drug administration implies that the nurse maintains professional curiosity, accountability, and integrity throughout her or his professional nursing career. This chapter focuses on the knowledge and skill the nurse must possess when using the nursing process as a framework for the safe administration of drug therapy. After completing this chapter the student will be able to:

- Define selected terms related to administration of medications.
- Describe legal aspects of administering drugs.
- Identify physiologic factors and individual variables affecting drug action.
- Describe various routes of drug administration.
- Identify essential parts of a drug order.
- Give examples of various types of medication orders.
- Recognize abbreviations commonly used in medication orders.
- List five essential steps to follow when administering drugs.
- Describe physiologic changes in elderly persons that alter drug administration and effectiveness.
- Outline steps required to administer oral medications safely.
- Identify equipment required for parenteral medications.
- Describe how to mix selected drugs from vials and ampules.
- Identify sites used for subcutaneous, intramuscular, and intradermal injections.
- Describe essential steps for safely administering parenteral medications by intradermal, subcutaneous, intramuscular, and intravenous routes.
- Describe essential steps in safely adminstering topical medications: dermatolologic, ophthalmic, otic, nasal, vaginal, and rectal preparations.

NAMES AND TYPES OF MEDICATIONS

1. ________________________ is a substance administered for the diagnosis, cure, treatment, relief, or prevention of disease.

2. ______________________ is the written direction for the preparation and administration of a drug .

3. ______________________ is the name under which a drug is listed in one of the official publications.

4. ______________________ is the name given by the drug manufacturer.

5. ______________________ is the study of the effect of drugs on living organisms.

6. ______________________ is a person licensed to prepare and dispense drugs and to make up prescriptions.

DRUG STANDARDS AND LEGAL ASPECTS

Describe the type of information about drugs found in the following books.

7. Pharmacopeia ______________________

8. United States Pharmacopeia ______________________

9. Formulary ______________________

What does the nurse need to know to protect himself or herself legally when administering drugs?

10. ______________________

____

DRUG MISUSE

Define the following terms.

11. Drug abuse ______________________

12. Drug dependence ______________________

13. Physiologic dependence ______________________

14. Psychologic dependence ______________________

EFFECTS OF DRUGS

Match the terms in column I by drawing a line to the appropriate description in column II.

15. Therapeutic effect
16. Side-effect
17. Drug toxicity
18. Drug allergy
19. Anaphylactic reaction
20. Drug tolerance
21. Cumulative effect
22. Idiosyncratic effect
23. Drug interaction
24. Iatrogenic disease

A. A person who has unusually low physiologic activity in response to a drug

B. A severe allergic reaction occuring immediately after administration of the drug

C. A disease caused unintentionally by medical therapy

D. The primary effect intended or the reason the drug is prescribed

E. A deleterious effect of a drug on an organism or tissue

F. Occurs when administration of one drug alters the effect of another

G. An unintended secondary effect of a drug

H. The immunologic reaction to a drug tow which a person has already been sensitized

I. The increasing response to repeated doses of a drug that occurs when the rate of administration exceeds the rate of metabolism or excretion

J. An unexpected and individual over or underresponse to a drug

ACTIONS OF DRUGS ON THE BODY

Indicate the percent of drug in a client's body at the following time intervals if the half-life of the drug is 4 hours.

25. After 4 hours _____ %

26. After 8 hours _____ %

27. After 12 hours _____ %

28. After 16 hours _____ %

ROUTES OF ADMINISTRATION

Describe the various routes of drug administration as well as the advantages and disadvantages of each in the chart that follows.

ROUTE OF ADMINSTRATION	DESCRIPTION
29. Oral	
30. Sublingual	
31. Buccal	
32. Subcutaneous	
33. Intramuscular	
34. Intradermal	
35. Intravenous	

MEDICATION ORDERS

List the six essential parts of a drug order.

36. ______________________________

37. ______________________________

38. ______________________________

39. ______________________________

40. ______________________________

41. ______________________________

ADMINISTERING MEDICATIONS SAFELY

List the five "rights" of drug administration.

42. ______________________________

43. ______________________________

44. ______________________________

45. ______________________________

46. ______________________________

SELF ASSESSMENT QUESTIONS

47. While preparing a client's medications, the nurse notes that the medication sheet reads Demerol 75-100 mg q 4 h. The chart reads Demerol 100 mg q 3 - 4 h, p.r.n. What should the nurse give?

    A. Demerol 100 mg, q 4 h, p.r.n.
    B. Demerol 75 mg, q 3 h, p.r.n.
    C. Demerol 75 - 100 mg, q 4 h.
    D. Demerol 75 - 100 mg, q 3 - 4 h, p.r.n.

48. The doctor changes the demerol order to read: Demerol 50 - 75 mg, IM, q 3 - 4 h, p.r.n. The nurse interprets this order to mean

    A. alternate between 50 and 75 mg doses during the day.
    B. let the client determine how much of the drug he wishes to take.
    C. the nurse should determine which dose would be most effective.
    D. the nurse should call the doctor before deciding on the dose.

49. The nurse administered 75 mg of the Demerol after the client complained of pain. How should the medication be documented?

    A. Demerol 75 mg, IM.
    B. Dem. 75 mg, OD.
    C. Demerol 75 mg, PO.
    D. Demerol 75.

50. If a client complains of an upset stomach, generalized itching, and ringing in his ears, the nurse should

    A. discontinue all drugs and call the doctor.
    B. pump the client's stomach.
    C. withhold medications until further notice.
    D. ask the client if he is allergic to any drugs.

51. When administering medications the nurse should

    A. leave the medication at the bedside if the client is unavailable.
    B. readminister medications if the client vomits.
    C. follow the doctor's medication order with question.
    D. have another nurse check dosages of insulin and anticoagulants.

ADDITIONAL LEARNING ACTIVITIES

1. In a laboratory setting, get a classmate to play the role of the client. Determine the following injection sites:
   A. Right ventrogluteal (IM)
   B. Left dorsogluteal (IM)
   C. Left deltoid (IM)
   D. Right vastus lateralis (IM)
   E. Left arm (subcutaneous)
   F. Right back (intradermal)

2. In a clinical setting, list the medications ordered for two clients. Determine which orders are A) stat, B) standing, and C) p.r.n. Identify the agency medication policies that apply to these drug orders.

3. In a clinical setting, identify problems that clients of various age groups may have in taking oral medications. Identify the nursing measures that assist clients to take those oral medications.

4. Determine the different ways medications are administered intravenously in a clinical setting.

5. Interview two elderly clients taking daily medications. Assess their understanding of their medications.

46 WOUND CARE

Chapter 46 focuses on the nurse's role in promoting healthy wound healing. The processes of wound healing and related complications are discussed as well as the strategies used to implement care. This chapter focuses on the knowledge and skills the nurse needs to promote healthy wound healing. After completing this chapter the student will be able to:

- Define terms commonly used to describe wounds.
- State two basic ways in which wounds heal.
- Describe factors that affect wound healing.
- Identify the main complications of wound healing.
- Describe assessment criteria of a clean, healing wound.
- Identify nursing diagnoses for clients with various types of wounds.
- List suggested nursing strategies to promote wound healing and prevent complications of wound healing.
- Describe open and closed methods of wound care.
- Identify commonly used dressing materials and binders.
- Give reasons for selected steps of wound care procedures outlined in this chapter.
- Identify physiologic responses to heat and cold and purposes of heat and cold.
- Describe methods of applying dry and moist heat and cold.
- List outcome criteria by which to gauge whether wound healing has been achieved.

TYPES OF WOUNDS

Define the following terms.

1. Wound ______________________________

____

2. Intentional wound ______________________________

____

3. Unintentional wound______________________________

____

4. Incision ______________________________

____

5. Contusion ______________________________

____

6. Abrasions ______________________________

____

7. Puncture wound ______________________________

____

8. Laceration ______________________________

____

WOUND HEALING

Describe the three *internal* factors and three *external* factors that affect wound healing.

Internal factors:

9. ______________________________

10. ______________________________

11. ______________________________

External factors:

12. ______________________________

13. ______________________________

14. ______________________________

USING THE NURSING PROCESSS FOR WOUND HEALING

Complete the following table indicating at least four signs and symptoms of the complications of wound healing.

15. HEMORRHAGE	16. INFECTION	17. DEHISCENCE

What techniques should the nurse use when assessing a wound?

18. ____

____

Develop two nursing diagnoses for a client with a wound. Then develop at least one outcome criteria and one nursing intervention for each diagnoses.

NURSING DIAGNOSIS	CLIENT GOAL/ OUTCOME CRITERIA	NURSING INTERVENTION AND RATIONALE
19.		
20.		

SELF ASSESSMENT QUESTIONS

21. When a wound infection is suspected the nurse should

    A. reinforce the dressing.
    B. save a sample of drainage for the doctor.
    C. collect a specimen for culture and sensitivity.
    D. apply a binder to the area.

22. When attempting to control severe bleeding, the nurse should apply

    A. direct pressure.
    B. heat.
    C. friction.
    D. a bandage.

23. When bandaging a client's leg, the nurse should bandage the leg

    A. in a hyperextended position.
    B. from the proximal end to the distal end.
    C. leaving the toes exposed.
    D. using the greatest pressure over the calf.

24. Clean-contaminated wounds are

    A. uninfected wounds in which no inflammation is encountered.
    B. surgical wounds in which the respiratory, alimentary, genital or urinary tract has been entered.
    C. wounds in which there has been a major break in sterile technique.
    D. old, accidental wounds containing dead tissue or show evidence of clinical infection.

25. When giving a client a cooling sponge bath, the nurse should

    A. gradually decrease the temperature of the water.
    B. check the vital signs every 30 minutes.
    C. Omit the client's face and chest.
    D. place cold washcloths or ice packs in the axillae and groins.

ADDITIONAL LEARNING ACTIVITIES

1. Follow a nurse on rounds who specializes in the care of skin and ostomies. What kinds of treatments does the nurse use to promote healing and prevent complications?

2. Practice applying various types of bandages on a classmate in a practice laboratory.

47 PERIOPERATIVE CARE

Throughout the entire process of caring for the client undergoing surgical intervention, the nurse brings together myriad concepts aimed at providing comprehensive care resulting in the attainment of the client's highest level of health. During the preoperative phase the nurse focuses on the client's problems and needs related to the impending surgery. In this phase emphasis is place on preparation for diagnostic procedures, client teaching, and the legal aspects of informed consent. During the intraoperative period the surgical nurse focuses on safety measures to protect the client when he or she is most vulnerable. Complications from anesthesia administration or postsurgical problems such as a hemorrhage require a nurse with special skills in the immediate postoperative period. The recovery room nurse is skilled in anticipating untwoard effects and initiating appropriate interventions. This chapter brings together the concepts, knowledge, and skills that the nurse must possess to apply the nursing process to the client in each phase of perioperative care.

- Describe the phases of the perioperative period.
- Discuss the legal aspects of surgery.
- Describe the elements of surgical risk.
- Outline the various aspects of preoperative assessment.
- Give examples of pertinent nursing diagnoses for surgical clients.
- Identify the essential nursing responsibilities included in planning perioperative nursing care.
- Describe how to teach clients to move, perform leg exercises, and perform coughing and deep breathing exercises.
- Identify the essentials of preoperative skin preparation.
- Explain why gastric intubation may be used for surgical clients.
- Describe some of the ways to protect a client from injury.
- Identify circumstances in which the nurse monitors a client.
- Discuss the importance of documentation with reference to preoperative, intraoperative, and postoperative recording.
- Identify potential postoperative complications and describe nursing interventions to prevent them.
- Identify outcome criteria by which to evaluate the effectiveness of perioperative nursing interventions.

THE PERIOPERATIVE PERIOD

Define the the perioperative period and its three phases.

1. Perioperative period ____________________

2. Preoperative phase____________________

3. Intraoperative phase____________________

4. Postoperative phase____________________

Match the types of surgery in column I with the appropriate description in column II.

5. _____ Urgent
6. ______ Elective
7. ______ Optional
8. ______ Major
9. ______ Minor
10. ______ Diagnostic
11. ______ Exploratory
12. ______ Palliative
13. ______ Reconstructive
14. ______ Constructive
15. ______ Ablative

A. The repair of tissues or organs whose appearance or function has been damaged
B. Involves a high degree of risk and may be complicated
C. Essential but not necessarily an emergency procedure
D. Enables the surgeon to confirm a diagnosis
E. Usually requested by the client and not necessary for physical health
F. The removal of a diseased organ
G. Performed to relieve symptoms of a disease process
H. Performed for the client's well-being but is not necessary for life
I. Involves little risk and produces few complications
J. Performed to determine the extent of a pathologic process
K. Performed to correct a congenitally malformed organ or tissue

USING THE NURSING PROCESS: PREOPERATIVE PHASE

List and discuss the six factors that contribute to the degree of risk involved in a surgical procedure.

16. ____

____

17. ____

____

18. ____

____

19. ____

____

20. ____

____

21. ____

____

Develop two nursing diagnoses for the client in the preoperative period. Then develop at least one outcome criteria and one nursing intervention for each diagnosis.

NURSING DIAGNOSIS	CLIENT GOAL/ OUTCOME CRITERIA	NURSING INTERVENTION AND RATIONALE
22.		
23.		

INTRAOPERATIVE PHASE

Describe the two roles nurses perform in the intraoperative phase.

24. ______________________________

25. ______________________________

POSTOPERATIVE PHASE

What areas of client functions are assessed in the recovery room?

26. ______________________________

USING THE NURSING PROCESS: POSTOPERATIVE PHASE

Develop two nursing diagnoses for the client in the postoperative period. Then develop at least one outcome criteria and one nursing intervention for each diagnosis.

NURSING DIAGNOSIS	CLIENT GOAL/ OUTCOME CRITERIA	NURSING INTERVENTION AND RATIONALE
27.		
28.		

Describe the following potential postoperative complications, the clinical signs and preventive interventions.

PROBLEM AND DESCRIPTION	CLINICAL SIGNS	PREVENTIVE INTERVENTIONS
29. Atelectasis		
30. Hemorrhage		
31. Urinary Infection		
32. Wound dehiscence		

SELF ASSESSMENT QUESTIONS

33. Surgery is categorized as ablative when it

    A. enables the surgeon to confirm a diagnosis.
    B. is performed to relieve symptoms.
    C. repairs tissues that are damaged.
    D. removes diseased tissue.

34. Which of the following is *least* effective in the prevention of thrombophlebitis?

    A. leg exercises
    B. leg massage
    C. antiembolic stockings
    D. ambulation

35. A client who takes anticoagulant drugs before surgery should be observed for possible

    A. hemorrhage.
    B. hypotension.
    C. electrolyte imbalance.
    D. anaphylactic shock.

36. When a client verbalizes fear of death during surgery, the most important responsibility of the nurse is to

    A. reassure the client.
    B. give a sedative.
    C. make a psychiatric referral.
    D. notify the surgeon.

37. When preparing the client's skin for surgery the nurse should never

    A. use a depilatory.
    B. apply pressure when using clippers.
    C. stretch the skin when using a razor.
    D. use a disinfectant after removing hair.

ADDITIONAL LEARNING ACTIVITIES

1. Follow a client who is receiving surgery through the preoperative, intraoperative and postoperative phases. Carry out as many nursing activities as possible such as preoperative preparation and teaching. Accompany the client to surgery and observe the surgical procedure if possible. Accompany the client to the recovery area. How did the nurses in each area protect and provide safety for the client?

48 SPECIAL STUDIES

Clients who are receiving diagnostic tests or treatments are often anxious and fearful of either the pain associated with the test or the outcome of the procedure. The nurse's responsibility in caring for these clients is to be aware of their psychologic and physiologic needs before, during, and after the procedures are completed. Chapter 48 focuses on the knowledge and skills the nurse needs to meet the needs of the client undergoing special studies. After completing this chapter the student will be able to:

- Define terms related to selected special procedures used in diagnosing and treating clients.
- Describe the purposes and sequencing of selected procedures.
- Identify assessment data required for specific procedures.
- Identify education needs of clients about to have certain tests and treatments.
- Outline measures to prepare the client physically for specific procedures.
- Identify the nursing responsibilities during selected procedures.
- Describe guidelines for evaluating and recording client responses after special procedures.

GENERAL NURSING GUIDELINES

Write a short paragraph describing the responsibilities of the nurse in caring for a client receiving a diagnostic test and treatments before, during, and after the completion of the procedure.

1. ____

____

____

____

____

____

EXAMINATIONS

Briefly describe the following procedures and the three most important nursing interventions required during and after the procedure in the table below.

EXAMINATION AND DESCRIPTION	NURSING INTERVENTIONS
2. Bronchoscopy	
3. Cystoscopy	
4. Proctoscopy	
5. Gastrointestinal series	
6. Intravenous pyleography	
7. Cholecystography	
8. Myelography	

SELF ASSESSMENT QUESTIONS

9. After an electroecardiograph the client should

   A. remain flat in bed for 24 hours.
   B. return to normal activity.
   C. take a laxative.
   D. increase fluid intake.

10. When preparing a client for an electromyogram the nurse should tell the client that

    A. the procedure is painless.
    B. there may be some discomfort.
    C. anesthesia will be necessary.
    D. narcotics will be administered.

11. Immediately after a laryngoscopy the nurse should

    A. withhold food and fluids.
    B. maintain the client on bedrest.
    C. reduce the light in the room.
    D. sedate the client.

12. During a colonoscopy the nurse maintains the client in a

    A. supine position.
    B. fowler's position.
    C. Lithotomy position.
    D. Knee-chest position.

13. When assisting the physician perform a spinal tap the nurse should position the client

    A. with the legs elevated in stirrups.
    B. laterally with the head bent toward the chest, the knees flexed onto the abdomen.
    C. in a prone position with the head turned toward the nurse.
    D. with the back hYperextended and the arms abducted.

ADDITIONAL LEARNING ACTIVITIES

1. Prepare and implement a teaching plan for a client who is to undergo a diagnostic procedure. Accompany and observe the client undergoing the test. Did the preexamination teaching that the client receive, relieve anxiety and stress related to the test?

2. Spend a morning in the x-ray department observing various types of procedures being performed. How will your nursing care of clients undergoing x-ray examination differ after making these observations?

3. Observe a nurse caring for a client undergoing a proctoscopy, sigmoidoscopy, or colonoscopy. How did the nurse help the client relieve anxiety during the procedure?

APPENDIX
ANSWER SECTION

CHAPTER 1

1. Caring
2. Adaptive
3. Individualized
4. Holistic
5. Family and community interrelated
6. National League of Nursing Education
7. Isabel Hampton Robb
8. Florence Nightingale
9. Mary Mahoney
10. Rockefeller Survey (Committee for the Study of Nursing Education)
11. Profession
12. Profession
13. Profession
14. Non profession
15. Profession
16. Non profession
17. Gains a body of knowledge in a university setting
18. Attains competencies derived from a theoretical basis
19. Delineates and specifies the skills and competencies that are boundaries of expertise
20. Assesses, plans, implements, and evaluates theory, research, and practice in nursing
21. Accepts, promotes, and maintains the interdependence of theory research, and practice
22. Communicates and disseminates theoretical knowledge, practical knowledge, and research findings to the nursing community
23. Upholds the service orientation of nursing in the eyes of the public
24. Preserves and promotes the professional organization as the major referent
25. Professional definition and regulation
26. Development of a knowledge base for practice
27. Transmission of values, norms, knowledge, and skills
28. Communication and advocacy of the values and contributions of the field
29. Attendance to the social and general welfare of their members
30. Health promotion
31. Health maintenance
32. Health restoration
33. Care of the sick and dying
34. Home health
35. Community agencies
36. Ambulatory clinics
37. Health maintenance organizations
38. Nursing practice centers

39. J	46. B	52. T
40. G	47. L	53. T
41. M	48. E	54. F
42. C	49. A	55. F
43. I	50. K	56. F
44. H	51. F	57. T
45. D		

58. American Academy of Nursing
59. National Student Nurses' Association
60. International Council of Nurses
61. World Health Organization
62. Sigma Theta Tau
63. National League For Nursing
64. Canadian Nurses' Association
65. American Nurses' Association

66. B	69. A
67. C	70. B
68. A	

CHAPTER 1 CROSSWORD

ACROSS

2. Lillian Wald
5. Care of dying
7. ND
8. MS
11. clinician
12. Codes
13. NE
14. IM
15. Image
17. Eye
21. ANA
22. Sigma Theta Tau
24. M. Mahoney
25. Robb
27. MA

28. NCLEX
29. SN
30. RNCS
31. PhD
34. NP
36. Foleys
37. V. Henderson
39. American Nurses Association
42. LVN
44. WHO
45. Consumer
47. HMO
48. ND
49. ANN
50. Client
51. Nightingale
52. Florence
56. Profession
58. Breckenridge
59. MD
60. Occupation
62. Autonomy
64. Mary Brewster
66. Eye
67. ACT
68. Augustinian sisters

DOWN

1. FNS
2. Linda Richards
3. National League for Nursing
4. AAN
5. Case
6. MSN
8. Mary Nutting
9. MSN
10. FAAN
12. CNA
13. Nurse Theorist
14. I. Hampton
15. International Council of Nurses
16. Get
18. Mary Snively
19. Standard
20. Hampton
23. Ethics
26. BSN
32. His
33. Case method
35. DNS
40. Mary
41. Diploma
43. Vocation
45. Career
46. NLN
47. HS
49. All
53. Promotion
54. Need
55. Primary
56. Pro
57. NSNA
61. DNSC
63. MS
65. BSN

CHAPTER 2

1. People learn to become members of society
2. Early socialization - birth to adolescence
3. Mutual learning between peers, children and parents, or students and teachers
4. Learning to adjust to new situations
5. People prepare themselves for roles to which they aspire but which they do not yet occupy.
6. Learning a different way of looking at the world and the process of changing behavior in rather dramatic ways.
7. Values and norms of the profession are internalized into one's own behavior.
8. Learning and adapting to new roles
9. The development of a professional soul
10. A strong commitment to the service that nursing provides for the public.
11. Belief in the dignity and worth of each person
12. A commitment to education
13. Autonomy
14. Proficiency in specific work tasks
15. Attachment to significant others in the work or reference group
16. Internalizes the values of the professional group and adopts the prescribed behaviors
17. Initial innocence
18. Labeled recognition of incongruity
19. "Psyching out"
20. Role simulation
21. Provisional internalization
22. Stable internalizatin
23. Transition of anticipatory role expectations of societal group
24. Attachment to significant others
25. Internalization of role values/behaviors
26. Novice
27. Advanced beginner
28. Competent
29. Proficient
30. Expert
31. Dalton- performance of routine duties of the new graduate under the direction of a mentor. Kramer - mastery or development of technical expertise.
32. Dalton- independence develops at this stage. Kramer -social integration occurs.
33. Dalton-behavior is influencing, guiding, directing, and developing others.Kramer - behavior includes moral outrage at incongruities between the bureaucratic, professional and service roles.
34. Dalton-influence is exerted on the organization. Kramer-conflict resolution by surrendering behaviors and/or values occurs.

35. D
36. A
37. E
38. C
39. B
40. A
41. B
42. B
43. C
44. A

CHAPTER 2 CROSSWORD

ACROSS

1. Health
4. Situational
7. M .E. (ab)
8. Advocate
12. Ears
14. Care
15. Cue
16. RN
17. C.U. (ab)
18. Lead
19. ANA
20. Laissez-Faire
22. Pri
23. Delegation
25. CNA
28. Primary
29. Change
31. Helps
33. cence
34. Authority
36. Value
39. Expert
41. Researcher
45. Reciprocal
46. Role

DOWN

1. Helper
2. am
3. Leads
4. Socialization
5. Ounce
6. Autocratic
9. Values
10. Leadership
11. Secondary
13. AN
21. Agent
24. ER
27. Behavior
29. CS
30. Adult
32. ER
35. initial
37. UN
38. Exit
40. AS
42. Eye
43. Cap
44. DL

CHAPTER 3

1. Identify symptoms
2. Diagnose the problem
3. Explore alternative solutions
4. Select one course of action
5. Plan the steps in the change process
6. Implement the change
7. Evaluate the outcomes
8. Refreeze the client system
9. Empiric- rational
10. Normative-reeducative
11. Power-coercive
12. More acutely ill clients in hospitals, more home health nursing; more outpatient serices being used by low income population require nurses to identify how their knowledge and skills can fit into these settings.
13. AMA has attempted to create a new category of health care worker, the registered care technologist which nurses fear will fragment nursing care. Employers and educators of nurses are testing strategies and programs that will address the shortage problem.
14. Increasing involvement in the general public in health care issues have led nursing associations and regulatory agencies to have consumer representation on their governing boards.
15. New family structures are requiring nursing to provide better and different nursing care services.
16. Technology has decreased the need for special nursing care and has required nurses to become highly specialized.
17. Legislation affects how client's health needs are funded and met indirectly affecting nursing practice.
18. Established the need for, type of and quality of nursing care needed based on statistical data.
19. Has influenced members of the nursing profession to be more assertive regarding professional rights and autonomy
20. Collective bargaining improves and maintains quality in the profession.
21. Nursing association accreditation programs continually upgrade standards of nursing education programs. Credentialing programs have given status to expanded nursing roles. Political action branches of nursing associations influence health and nursing legislation.

22. T
23. F
24. T
25. F
26. F
27. F
28. T

29. American Journal of Nursing
30. Image: Journal of Nursing Scholarship
31. Advances in Nursing Science
32. The Annual Review of Nursing Research
33. Heart and Lung
34. Information concerning the nature of the study; and harm that could result from being in the study and her rights and responsibilities.

35. "Your care will be identical in quality whether or not you agree to participate."
36. Right not to be harmed.
37. Right to full disclosure.
38. Right to self determination.
39. Right of privacy and confidentiality
40. Responsibility to protect the client's rights as a subject in a research study
41. May lead to a new and better approach in treating clients with leg ulcers.
42. Define the purpose of the study.
43. Review related literature.
44. Formulate hypotheses and defining variables.
45. Select the research design.
46. Select the population, sample, and setting.
47. Conduct a pilot study.
48. Collect the data.
49. Analyze the data.
50. Communicate conclusions and implications.
51. Automated client care planning systems
52. Expert systems
53. Research assistance
54. Administrative assistance
55. Power
56. Coerce
57. Referent
58. Connection
59. Expert
60. Action
61. Government
62. ANA
63. CNA
64. NLN
65. Community
66. Influence
67. Nurses have the ability and opportunity to see the changes that are needed to upgrade client care, the health care delivery system and the profession.

68. - 70. Should include 3 of the following: Identify symptoms, diagnose the problem, explore alternative solutions, select action, plan the steps of the change process, implement measures, evaluate outcomes and refreeze the system.

71. B
72. B
73. D
74. B
75. B

CHAPTER 4

1. Assessing
2. Diagnosing
3. Planning
4. Implementing
5. Evaluating

6. E
7. G
8. C
9. B
10. I
11. H
12. A
13. D

14. Concept
15. Nursing process
16. Scientific knowledge
17. Deductive
18. Inductive
19. Level of abstraction
20. Nursing research
21. The goal of nursing is what nursing is trying to achieve. Each nursing model specifies its own goals.
22. The tern client refers to the recipient of nursing service.
23. Role of the nurse must be wanted, needed and accepted by society. It is described differently in various nursing models
24. The probable origin or cause of any client problems amenable to nursing intervention
25. The target of nursing intervention or the concentration of nursing care
26. Spell out the specific ways in which the nurse helps the client
27. Reflects the nursing goal and the concept of the client.
28. Open system
29. Environment
30. Subsystems
31. Biopsychosocial
32. Subsystem
33. Whole
34. See pg. 74 and Figure 4-2 in *Fundamentals of Nursing, fourth edition*
35. Self actualization
36. Self-esteem needs
37. Love and belonging needs
38. Safety and security needs
39. Physiologic needs
40. Forming a humanistic-altruistic system of values
41. Instilling faith and hope

42. Cultivating sensitivity to one's self and others
43. Developing a helping-trust relationship
44. Expressing positive and negative feelings
45. Using a creative problem-solving caring process
46. Promoting interpersonal teaching-learning
47. Providing a supportive, protective, or corrective mental, physical, sociocultural, and spiritual environment.
48. Assisting with the gratification of human needs.
49. Allowing for existential-phenomenologic-spiritual forces.
50. Trial-and-error
51. Intuition
52. Experimentation
53. Scientific method
54. Modified scientific method
55. Deliberation
56. Judgment
57. Discrimination
58. Helps nurses to communicate and collect as well as interpret data about clients.
59. Dorothy Johnson
60. Systems theory
61. Imogene King
62. Framework
63. Conceptual framework
64. Diagnosing
65. System
66. Assessing
67. Self actualization
68. Humanism
69. Deductive research
70. Virginia Henderson
71. Need
72. Maslow
73. Martha Rogers
74. Holism
75. Watson
76. Planning
77. Inductive research
78. Dorothea Orem
79. Theory
80. Value system
81. Betty Neuman
82. Myra Levine
83. Concept
84. D
85. A
86. A
87. D
88. D

CHAPTER 5

1. B
2. C
3. A

4. - 8. Includes any of the following: Genetic or family predisposition, environmental abnormalities, biologic agents, physical agents, chemical agents, faulty production of antibodies, continued stress, age, nutritional status and occupation.

9. The ability to conceptualize a state of health and respond to changes in health are directly related to age.
10. Concepts about health are often transmitted from parents to children.
11. Previous experiences with health and illness often determines how a person perceives health in the present.
12. Any variation from the level of functioning a person expects in himself may affect the perception of health.
13. Any threat to how a person perceives the self may result in anxiety. This may cause the person to reassess health definitions and status.
14. A family history of a certain disorder may make an individual feel at high risk.
15. The client believes the illness may cause death or have serious consequences.
16. The combination of perceived susceptibiity and perceived seriousness determine the total perceived threat of an illness to a specific individual.
17. A client may reject a nurse's suggestions and interventions and become angry because of intrusion into personal habits.
18. Compliance is the extent to which a person's behavior coincides with health practitioners' advice.
19. The client's motivation to become well.
20. The value a client places on reducing the threat of illness.
21. The client's belief that compliance will reduce that threat.
22. Demonstrate caring
23. Encourage health behaviors through positive reinforcement.

24. Establish why the client is not following the regimen.
25. Use aids to reinforce teaching.
26. Establish a therapeutic relationship of freedom, and mutual responsibility with the client and support persons.
27. Symptom experience stage
28. Transition stage
29. Physical experience of symptoms
30. Cognitive aspect
31. Emotional response
32. Acceptance
33. Changed
34. Better
35. Treatment
36. Medical care contact
37. Validation
38. Explanation
39. Reassurance
40. Dependent patient role
41. Independence
42. Accepting
43. Environment
44. Recovery or rehabilitative
45. Sick role
46. Roles and functions.
47. The activity undertaken by those who consider themselves ill, for the purpose of getting well.
48. The sick person is excused from some normal duties.
49. Clients whose prognosis is poor or uncertain are permitted more dependence than people who are less seriously ill.
50. The person who fears dependence may be threatened by assuming a sick role and may ignore advice despite the consequences.
51. Some clients find dependence so satisfying that they perptuate the sick role and do not try to get well or continue to complain of symptoms even after they are well.
52. Ascertain what privacy means to the client and try to support the client's needs whenever possible.
53. Allow clients to have as much input into the plan of care as possible. Learning and individualizing the plan of care is also important.
54. Nurses can help clients adapt to a hospital by making arrangements to follow the client's routines at home, providing explanations about hospital routines, giving information to other personnel about the client's lifestyle when appropriate, reinforcing lifestyle changes in the hospital that can become permanent at home.
55. Provide care that is as economical and as safe as possible. Referrals should be made to social service departments to assist clients with financial problems.
56. Information regarding role changes, task reassignments, increased stress due to anxiety, financial problems, loneliness, change in social customs and the problems associated with caring for an elderly relative.

57. T
58. F
59. T
60. F
61. T
62. F
63. T
64. F
65. T
66. F
67. F
68. F
69. T
70. F
71. F
72. T
73. F
74. T
75. F
76. F
77. F

78. Health promotion and protection against specific illnesses
79. Protection and maintenance of health and prevention of complications or disabilities
80. Rehabilitation, restoration and prevention of further disability after an illness
81. B
82. A
83. D
84. D
85. C
86. A

CHAPTER 6

1. The totality of services offered by all health disciplines.

2. T
3. F
4. F
5. T
6. T
7. T
8. T
9. T
10. F
11. T

12. Have health care needs met at one time at one agency

13. Health care that reflects the holistic view of the total person
14. More information and services related to health promotion and illness prevention.
15. Social insurance is financed through federal funds. Voluntary insurance includes private insurance plans that are financed by the individual or by the individual and/or the employer.
16. Medicare - a national and state insurance for the aged over 65.
17. Medicaid - a federal public assistance program paid out of general taxes to people who require financial assistance.
18. Supplemental Security Income - special payments for the disabled or blind.
19. Workman's Compensation - A fund supported by employer companies to make payments to injured workers.
20. Tertiary - promoting, maintaining and restoring health in the client's home.
21. Primary - emergency services for clients experiencing life crises. Also guidance, counseling and support for long term therapy.
22. Secondary - emergency care restorative.
23. Tertiary - skilled nursing facilities for extended care, intermediate care and personal care for the chronically ill or disabled.
24. Primary - basic and supplemental health maintenance and treatment services to voluntary enrollees.
25. Tertiary - a variety of services to the terminally ill and their families. Supports palliative measures for relief rather than cure.
26. Primary - medical, nursing, laboratory, radiologic, and minor surgical services.
27. Secondary - restorative care to the ill or injured. Also health related research and teaching.
28. The psychomotor, affective and the cognitive skills along with activities associated with the nurse's expanded role.
29. A person who is licensed to practice medicine in a particular jurisdiction.
30. Prepares and dispenses pharmaceuticals in hospital and community settings. Monitors and evaluates the actions and effects of medications on clients
31. Counsels clients and support persons about social problems, such as finances, marital difficulties, and adoption of children.
32. Has special knowledge about nutrition and food.
33. Assists clients with musculoskeletal problems with heat, water, exercise, massage, and electric current.
34. Considerate and respectful care.
35. Complete current information concerning the diagnosis, treatment, and prognosis from the physicain.
36. Information necessary to give informed consent prior to the start of any procedure and/or treatment.
37. To refuse treatment and to be informed of the medical consequences of his action.
38. Consideration of privacy concerning the medical care program.
39. All communications and records pertaining to care should be treated as confidential.
40. Reasonable response to the request of a patient for services.
41. Information as to any relationship of the hospital to other health care and educational institutions insofar as the patient's care is concerned.
42. To be advised if the hospital proposes to engage in or perform human experimentation affecting the patient's treatment, and the right to refuse treatment.
43. Reasonable continuity of care.
44. Explanation of his bill regardless of source of payment.
45. Right to know what hospital rules and regulations apply to his conduct as a patient.
46. Means receiving care from 5 to 30 people during a client's hsopital experience. This creates problems with the smooth flow of information and plans to help the client.
47. Need for newer and better methods of treatment and equipment, need for additional space, inflation, increased population demands, and increased number of health-illness care providers.
48. There are 2 to 3 million homeless in the United States. The numbers are increasing and drug abuse, deinstitutionalization of mental health facilities.

49. Elderly experience long-term illness and disablity which require special needs such as housing, treatment services and financial support.
50. Health support services and personnel are unevenly distributed throughout the U.S. and Canada with largest number in highly urbanized areas.
51. Community-based health systems in which primary care is the major function.
52. Redistribution of health and specialty services to overcome regional inequities.
53. Self-reliance and participation by community in health matters.
54. Provision of services to specific target groups.
55. Increased involvement of existing health organzations and groups.
    See Chapter 6 for additional priorities.
56. Clarify nursing knowledge. The title "nurse," should be restricted to the professional nurse.
57. High standards of intelligence and motivation for those entering the profession
58. Remodeling of the nursing education system to include more indepth sciences, economics, legal and ethical issues, management and business, information technology, and greater clinical application.
59. Preparation in self-governance and self-management
60. Increased political action.
61. D
62. A
63. C
64. B
65. A

CHAPTER 7

1. B
2. D
3. C
4. F
5. E
6. B
7. Ethics concerning life
8. Behavior which involves judgements, actions, and attitudes based established norms.
9. The cognitive and developmental process of reasoning about moral choice.
10. The group of values a person holds.
11. - 13. Should include at least 3 of the following: Association with and observation of behavior and attitudes of parents and teachers. In addition, all life experiences such as interacting with cultural, religious and social environments.
14. Modeling
15. Responsible choice
16. Laissez-faire
17. Moralizing
18. Individuals find their own answers (values) to situations.
19. Prizing and cherishing
20. Publicly affirming when appropriate
21. Choosing from alternatives
22. Choosing after consideration of consequences
23. Choosing freely
24. Acting
25. Acting with a pattern, consistency, and repetition.
26. Publicly affirming when appropriate
27. Acting with pattern, consistency and repetition
28. Choosing after consideration of consequences
29. Prizing and cherishing
30. Helping a client discover a new and meaningul value system following illness or injury.
31. Providing information about the clien'ts respones to injury or illness
32. Helping the client explore alternative goals and intervention strategies when valued goals cannot be realized
33. Planning nursing interventions that support the client's cultural and health care beliefs.
34. Give Mrs Shultz as much information as possible so that she will be able to consider all courses of action. Help her to understand and examine the consequences of each course of treatment by discussing the consequences with her. Make sure decisons are made without pressure. decision to make certain she feels comfortable with her decision and is willing and able to discuss it with others. Give her support to carry out her decision.

35. Ethics refers to publicly stated and formal sets of rules or values, while morals are values or principles to which one is personally committed.
36. The client is at significant risk of harm, loss, or damage if the nurse does not assist.
37. The nurse's intervention or care is directly relevant to preventing harm.
38. The nurse's harm will probably prevent harm, loss, or damage to the client.
39. The benefit the client will gain outweighs any harm the nurse might incur and does not present more than minimal risk to the health care provider.
40. Awareness of different options.
41. Moral nature of the dilemma.
42. Two or more options with true choice.
43. Establish a sound database.
44. Identify the conflicts presented by the situation
45. Outline alternative actions to the proposed course of action
46. Outline the outcomes or consequences of the alternative actions.
47. Determine ownership of the problem and the appropriate decision maker
48. Define the nurse's obligations.
49. Client will die feeling that no one cared about his wishes. Joyce will feel guilty because she did not act as an advocate for the client.
50. Joyce will pass the responsibility for carrying out the client's wishes to her immediate supervisor who may be able to communicate the client's wishes more effectively to the physician. Joyce may feel she has done all that she could do in this situation.
51. Joyce may, or may not, sensitize the doctor to the client's needs, and in doing so, risks the chance of creating problems between herself and the doctor.
52. Clients who feel that nothing more can be done may feel hopeless, powerless, and lose the ability to cope with the disease.
53. A
54. A
55. B
56. A
57. D

CHAPTER 8

1. A system of principles and processes by which people, who live in a society, attempt to control human ocnduct in an effort to minimize the use of force as a means of resolving conflicting interests.
2. It provides a framework for establishing which nursing actions are legal.
3. It differentiates the nurse's responsibilities from those of other health professionals.
4. It helps to establish the boundaries of independent nursing action.
5. It assists in maintaining a standard of nursing practice by making nurses accountable under the law.
6. The body of law that deals with relationships between individuals and the government and governmental agencies.
7. A division of public law that deals with actions against the safety and welfare of the public.
8. Private or civil law deals with relationships between private individuals.
9. Involves the enforcement of agreements among private individuals or ethe payment of compensation for failure to fulfill the agreements.
10. Defines and enforces duties and rights among private individuals that are not based on contractual agreements.
11. Registration.
12. Accreditation.
13. Nurse Practice Acts.
14. Credentialing.
15. Licenses.
16. Certification.
17. Standards of care define what a reasonable and prudent professional nurse with similar preparation and experience would do in similar circumstances.
18. Nurse practice acts legally define and describe the scope of practice for the practitioner in that institution.
19. Agency policies define the scope of practice for the practitioner in that institution.
20. The fact contracted for, the nurse's emlpoyment, the duties to be performed

and the services provided must all be legal.

21. The parties to the contract must be of legal age.
22. There must be mutual agreement about the service to be contracted for.
23. There must be compensation for the service to be provided.
24. Actions taken by nurses can affect safety of people, therfore nurses faced with a strike must make an individual decision whether or not to cross the picket line.
25. Will
26. Informed consent
27. Euthanasia
28. Crime
29. Negligence
30. No code
31. Felony
32. False imprisonment
33. Assault
34. Slow code
35. Autopsy
36. Manslaughter
37. Slander
38. Inquest
39. Libel
40. Coroner
41. Misdemeanor
42. Organ donation
43. Defamation
44. Right to die
45. Tort
46. Identify the client's name, initials, and ID number.
47. Give the date, time, and place of the incident.
48. Describe the facts of the incident.
49. Identify all witnesses to the incident.
50. Identify any equipment by number and any medication by name and number.
51. Document any circumstance surrounding the incident.
52. Malpractice insurance protects the nurse financially in the even of lawsuits and provides legal assistance.
53. D
54. D
55. D
56. C
57. D

CHAPTER 9

1. T
2. F
3. T
4. T
5. F
6. T
7. F
8. T
9. To establish a database
10. Obtain health history, perform physical assessment, review records and literature, interview support persons, validate assessment data
11. To identify the client's health care needs and to prepare diagnostic statements
12. Organize data, compare data against standards, group data, identify gaps and inconsistencies, determine the clients' health problems, risks, and strengths, formulate nursing diagnosis statements.
13. To identify the client's goals and appropriate nursing interventions.
14. Set priorities, write evaluation goals and outcome criteria in collaboration with client, select nursing strategies, consult other health personnel, write nursing orders and care plan.
15. To carry out planned nursing interventions to help the client attain goals.
16. Reassess client, update database, review care plan, perform or delegate planned nursing strategies.
17. To determine the extent to which goals of nursing care have been achieved.
18. Collect data about the client's response, compare the client's response to evaluation criteria, analyze the reasons for the outcomes, modify the care plan.
19. The system is open, flexible, and dynamic.
20. It individualizes the approach to each client's particular needs.
21. It is planned.
22. It is goal directed.
23. It is flexible to meet the unique needs of client, family, or community.
24. It permits creativity for the nurse and client in devising ways to solve the stated health problem.
25. It is interpersonel. It requires the nurse to communicate directly and consistently with clients to meet their needs.
26. It is cyclical. Since all steps are interrelated- no absolute beginning or end.

27. Emphasizes feedback.
28. Universally applicable
29. Nursing care is planned, evaluated and reassessed to meet the unique needs of the individual, family or community.
30. The written care plan is accessible to all persons involved in the client's care.
31. Clients are helped to develop skills related to their health care.
32. The NLN requires graduates to be competent in the use of the nursing process. Licensure examinations are organized around nursing process activities.
33. Well-written care plans give nurses confidence that nursing interventions are based on correct identification of the client's problems.
34. Enhances the skill and expertise of the nurse
35. Fulfills the nurse's legal obligation to the client
36. Learning and implementing the nursing process is a basic requirement for professional nursing competence and fulfills standards of nursing practices
37. The JCAHO requires that nursing process be utilized and a written plan of care be available for each client.
38. Provides a framework for accountability and responsibility in nursing and maximizes accountability and responsibility for standards of care.

39. D
40. A
41. D
42. A
43. C

CHAPTER 10

1. Validates a diagnosis
2. Used before writing a nursing intervention or in obtaining information about a client's response to the nursing strategies
3. Determines the outcomes of the nursing strategies and to evaluate goal achievment
4. All the information about a client
5. Gathering information about a client's health status
6. The acceptance of assumptions as fact
7. OD, VD
8. OD
9. OD, VD
10. SD
11. OD
12. SD
13. SD
14. OD
15. OD
16. OD

17. Support persons who know the client well
18. Supplement, verify or convey information and are an important source when the client is young, unconscious or confused
19. Include care givers who can provide additional information
20. Previous or current information
21. A source of information about the client and his illness in the past and present
22. Present and past health and illness patterns, coping behavior, health practices, previous illnesses and allergies
23. Nursing and related literature
24. Comparative standards and norms, cultural and social health practices, spiritual beliefs, assessment data, nursing interventions and evaluation criteria, and information about medical diagnoses, treatment and prognoses
25. Gathering data by using the five senses
26. Planned communication with a purpose
27. A physical examination to obtain objective data for the assessment phase
28. Observing - collect objective data through the five senses
29. Looking at a rash, palpating a lump, smelling acetone odor
30. Interviewing - to gather information from the primary source, the client
31. Taking a health history
32. Examining -to gather objective information
33. Performing a head-to-toe examination
34. Closed
35. Closed
36. Open
37. Closed
38. Open
39. Closed
40. Schedule interviews when the client is comfortable and free of pain, and interruptions are few.
41. Adequate privacy, well-lighted, ventilated and free from noise, and movement

42. Informal arrangement with nurse at 45 angle from the client no further than 6 feet apart
43. Inspection
44. Auscultation
45. Palpation
46. Percussion

47. B
48. F
49. A
50. E
51. C
52. D
53. G
54. A
55. D
56. D
57. D
58. A

CHAPTER 11

1. T
2. T
3. T
4. T
5. F
6. T
7. T
8. T
9. T
10. F
11. M
12. M
13. N
14. N
15. N
16. M

17. Actual health problem exists in the present. Potential health problems may occur if nursing interventions are not instituted.
18. The act of interpreting collected data. It includes organizing, comparing, and clustering data.
19. Determining the client's health problems, health risks, and strengths
20. Formulating the nursing diagnosis
21. Problem - the diagnostic category label or title. It is a description of the client's health problem for which nursing therapy is given.
22. Etiology - the contributing factors. This Identifies one or more probable causes of the health problem and gives direction to the required nursing therapy.
23. Signs and symptoms - the defining characteristics or cluster of signs and symptoms.
24. Incorrectly stated
25. Correctly stated
26. Correctly stated
27. Incorrectly stated
28. Incorrectly stated
29. Correctly stated
30. Incorrectly stated
31. Correctly stated
32. Promotes professional accountability and autonomy by defining and describing the independent area of nursing practice.
33. Provide an effective vehicle for communication among nurses and other health care professionals
34. Provide an organizing principle for the building of meaningful research.
35. A
36. D
37. C
38. A
39. D

CHAPTER 12

1. The process of designing nursing strategies required to prevent, reduce, oreliminate those client health problems identified and validated during the diagnostic phase.
2. Family members
3. Support persons
4. Health professionals
5. Care givers
6. Data
7. Diagnostic statements
8. Nursing diagnosis
9. Priority setting
10. Client goal
11. Criterion
12. Planning
13. Extrapolating
14. Nursing orders
15. Nursing therapy
16. Maslow's hierarchy of needs
17. Behavorial objectives
18. Outcome criteria
19. Client goal
20. Nursing strategy
21. Brainstorm
22. Nursing care plan
23. Rationale
24. Kardex
25. Hypothesizing
26. Nursing order
27. Long term goal
28. Outcome criteria
29. Outcome criteria
30. Nursing care plan
31. Nursing strategies

32. Signature

33.	T	39.	Incorrect
34.	F	40.	Incorrect
35.	T	41.	Incorrect
36.	T	42.	Correct
37.	F	43.	Incorrect
38.	T	44.	Correct

45. Expresses a desire to give himself a partial bath within 3 days
46. Washes his face, neck, chest and arms within 5 days
47. Walks the length of the hall (20 feet) with no shortness of breath within 1 week
48. Climbs 10 stairs with no shortness of breath within 2 weeks
49. Assist the family and client think through their typical day and week and process the changes they can anticipate when the ill person is at home.
50. Referrals and other actions should be initiated before the day of discharge so that preparations can be made for the client's return
51. D
52. A
53. A
54. B
55. B

CHAPTER 13

1. Putting the nursing strategies listed in the nursing care plan into action: it is the nursing action taken to attain the desired outcome of the client's goals.
2. An activity that the nurse initiates as a result of the nurse's own knowledge and skills.
3. To be answerable
4. A set of classifications that are ordered and arranged on the basis of a single principle or consistent set of principles
5. Those activities carried out on the order of and under a physician's supervision.
6. Those activities performed either jointly with another member of the health care team or as a result of a joint decision by the nurse and a health care team member.
7. A written plan specifying the procedure to be followed in a particular situation
8. A written document about policies, rules, regulations, or orders regarding client care
9. Reassessing the client
10. Validating the nursing care plan
11. Determining the need for nursing assistance
12. Implementing the nursing strategies
13. Communicating the nursing actions
14. Assessing must take place whenever the nurse is in contact with the client.
15. Cognitive skills including problem solving, decision making, critical thinking and creativity are crucial to safe, intelligent nursing care.
16. Interpersonal skills are all the activities people use when communicating directly with one another.
17. Technical skill are "hands-on" skills such as manipulating equipment, giving injections or repositioning clients.
18. Directed thinking
19. Associative thinking
20. Critical thinking
21. Associative thinking
22. Directed thinking
23. Creative thinking
24. Based on scientific knowledge and nursing research
25. Must be understood by the nurse
26. Must be adapted to the individual
27. Should always be safe
28. Require teaching, supportive, and comfort components.
29. Should always be holistic.
30. Should respect the dignity of the client and enhance the client's self esteem
31. The client's active participation in implementing nursing actions should be encouraged as health permits.
32. B
33. D
34. D
35. D
36. D

CHAPTER 14

1. Identifying whether or to what degree a client's goals have been met
2. Evaluating is concurrent because the nurse evaluates while implementing.

3. Terminal-after completing the nursing activity, the nurse evaluates whether the client's goals have been met.
4. Identify outcome criteria used to evaluate the client's response to nursing care.
5. Collect data so that conclusions can be drawn about whether goals have been reached.
6. Judge goal achievement. Three possible outcomes are possible: the goal was met, the goal was partially met or the goal was not met.
7. Relate nursing actions to client outcomes
8. Examining the client's care plan to determine if goals have been met
9. Modifying the plan of care requires the nurse to change the data in the assessment column, revise the nursing diagnoses, revise the client's priorities, goals and outcome criteria and establish new nursing strategies.
10. Goal 1 - partially met
11. Goal 2 - not met
12. Goal 3 - met
13. The nurse needs to reexamine the client's database, nursing diagnostic statements, goal statements and nursing strategies
14. Quality assessment is an examination of services only.
15. Quality assurance implies that efforts are made to evaluate and ensure quality health care.
16. Both examine and evaluate services, however quality assurance ensures that quality will be maintained.
17. Defining and clarifying the nature of nursing
18. Deciding what approach to take
19. Developing standards and criteria
20. Testing the criteria
21. C
22. A
23. B
24. D
25. B

CHAPTER 15

1. Reviews pertinent knowledge, considers potential areas of concern, and develops plans for interaction.
2. Opens the relationship with introductions. Client expresses concerns for seeking help. The nurse clarifies the problem. The contract is structured and formulated.
3. Trust and and rapport is enhanced. The nurse assists client to explore thoughts, feelings and actions. The client develops skills of listening and gains personal insight. The nurse and client plan programs in view of the long and short term goals.
4. The nurse and client accept feelings of loss. The client accepts the end of the relationship without feelings of anxiety or dependence.
5. Recognizing limitations and seeking assistance as required
6. Instituting a relaxed attending attitude, listening, paraphrasing, and clarifying.
7. The nurse uses empathy, respect, genuineness, concreteness, self-disclosure, confrontation, decision-making and goal-setting skills. The client uses nondefensive listening, self-understanding and risk-taking skills.
8. The nurse uses summarizing skills. The client develops the ability to handle problems independently.
9. D
10. A
11. A
12. C
13. B
14. E
15. Verbal uses the spoken or written word, whereas nonverbal uses other methods such as gestures or facial expressions. Examples of verbal are a speech, lecture or conversation. Examples of nonverbal are a frown, or thumbs up gesture.
16. Simplicity
17. Clarity
18. Timing and relevance
19. Adaptability
20. Credibility
21. Physical contact to 1 1/2 feet
22. 1 1/2 feet - 4 feet
23. 4 - 12 feet
24. 12 feet and beyond
25. Noises infants make

26. The repetition of sounds just spoken by another, by an older infant.
27. One word represents a whole sentence. Used by 12 - 18 months.
28. Self-centered, noncommunicative speech used by children from 4 - 11 years
29. The exchange of thoughts with others occurs in school-age children.
30. Language skills presented by the client
31. Adequacy of the language skills.
32. The chief method of communicating
33. Obstacles to language development
34. Specific forms of language impairment
35. Gestures used by the individual
36. Posture and facial expressions employed
37. Use of touch as a means of communication
38. The interpersonal distance with which a client feels comfortable
39. Grooming and appearance
40. Vocabulary of the client
41. Use of symbols and gestures
42. Hostility, aggression, assertiveness, reticence, hesitance, anxiety, or loquaciousness
43. Difficulties with verbal communication
44. Refusal or inability to speak
45. Impaired verbal communication related to anxiety and fear of hospitalization as evidenced by anger, hostility, refusal to express needs, and inappropriate uncooperative behavior.
46. Respond with straight forward, simple honest statements to provide reality orientation and correct faulty perceptions.
47. Acknowledge Mr. Jarguzis's feelings. Encourage him to express his fear and anxiety about his illness.
48. Help him look at the effects of his nonfacilitative communication patterns.
49. Gives clear , concise, understandable messages.
50. Expresses feelings.
51. Establishes a method of communication in which needs can be expressed.
52. Information giving. Should describe her role. Good use of personal space (+)
53. Open question, general lead (+). Nurse's body language may imply lack of interest, impatience or both (-).
54. Asking for clarification and reflects client's feelings (+). Asks closed question (-). Body language open (-).
55. Another closed question (-).
56. Body language (-).
57. Reflecting feeling (+).
58. Acknowledging feelings (+).
59. Paraphasing (+).
60. Defensive and closed body language (-).
61. Defensive, body language implies end of interview (-).
62. Responding and use of touch (+).
63. Accomplish its goals
64. Maintain its cohesion
65. Develop and modify its structure to improve its effectiveness
66. A
67. A
68. C
69. A
70. B

CHAPTER 16

1. E
2. G
3. C
4. A
5. H
6. D
7. B
8. F

9. Motivation to learn is the desire to learn. The term describes forces acting on or from within the person to initiate, direct, and maintain behavior.
10. Readiness to learn is the behavior that reflects motivation at a specific time. It sometimes comes with time and the nurse's role is to encourage its development.
11. Feedback is information relating a person's performance to a desire goal.
12. Elevated anxiety can impede learning. People who are anxious may not hear spoken words or may retain only part of the communication.
13. Physiologic events such as pain, or illness can impair learning because the client cannot concentrate.
14. Cultural barriers such as language or value differences impairs learning.
15. The cognitive domain includes intellectual skills such as thinking, knowing and understanding.
16. The affective domain includes feelings, emotions, interests, attitudes, and appreciations.

17. The psychomotor domain includes motor skills such as giving an injection.
18. Teaching activities should help the learner meet individual learning objectives.
19. Rapport between teacher and learner is essential.
20. Use the client's previous learnings to facilitate learning in the present.
21. Communicate clearly and concisely.
22. The teaching activities need to be oriented around the learning objectives.
23. The nurse can facilitate readiness by calling attention to his need to lose weight, giving him information to read and by pointing out opportunities to learn. Motivation can be facilitated by relating the learning to something Bob values, by helping him make the learning situation pleasant and by encoouraging self-direction.
24. Knowledge deficit: low-calorie diet related to newly prescribed therapy as evidenced by the client's obesity and refusal to discuss diet plan.
25. Will loose 25 pounds in six months
26. Will select foods from the hospital menu which are appropriate for a 1200 calorie diet within three days
27. Client contracting
28. Behavior modification
29. The optimal time for each session depends largely on the learner
30. The pace of each teaching session also affects learning.
31. An environment can detract or assist learning.
32. Teaching aids can foster learning.
33. Learning is more effective when the learners discover the content for themselves.
34. A
35. A
36. B
37. B
38. D

CHAPTER 17

1. Helps coordinate care given by several people.
2. Prevents the client from having to repeat information to each health team member.
3. Promotes accuracy in the provision of care and lessens the possibility of error.
4. Helps health personnel make the best use of their time
5. Communication
6. Legal documentation
7. Research
8. Statistics
9. Education
10. Audit
11. Planning client care
12. Admission sheet
13. Physician's order sheet
14. Medical history sheet
15. Nurse's notes
16. Special records and reports
17. Defined database
18. Problem list
19. Initial list of orders or care plans
20. Progress notes
21. Each person or department makes notations in a separate section of the chart.
22. Data about the client are recorded and arranged according to the problems the client has, rather than according to the source of the information.
23. A method of organizing and recording data about the client, consisting of a series of cards kept in a portable index file.
24. Used to create customized care plans, retrieve data, and provide work lists
25. A descriptive record of the client's progress which includes assessments, independent/dependent nursing interventions, evaluations, physician procedures and visits by health team members.
26. A graphic record of client variables such as pulse, blood pressure, medications, and progress in learning a new skill.
27. Notes completed when a client is being discharged and transferred to another institution or home.
28. C
29. D
30. A
31. B
32. Date and time for first entry are incomplete. There is no date or time for

second entry. Leg is spelled incorrectly and should be corrected by drawing line through the mistake. The word "error" should be written above it. A line should be drawn through the blank spaces. There is no signature for the second entry.

33. Abdomen
34. Complains of
35. Drops
36. At bedtime
37. Out of bed
38. Tincture
39. Female
40. Male
41. Times
42. Number, fracture
43. Less than
44. Yes. Hospital personnel may not release any information concerning a client's condition without written permission from the client. In some areas, a client who has been involved in an accident may be identified after the family has been notified.
45. The conferance allows each nurse to offer an opinon about the problem.
46. Rounds allows the client to participate in the discussion and allows.
47. D
48. D
49. B
50. B
51. C

CHAPTER 18

1. Cardinal
2. Functions
3. History
4. Pulse
5. Temperature
6. Respirations
7. Blood pressure
8. Basal metabolic rate (BMR)
9. Muscle activity
10. Thyroxine output
11. Epinephrine, norepinephrine, and sympathetic stimulation.
12. Increased temperature of body cells (fever)

13. A
14. E
15. B
16. C
17. D
18. G
19. F
20. 97.6°F
21. 102.8°F
22. 99.4°F
23. F
24. T
25. T
26. T
27. T
28. T
29. F
30. T
31. F

32. Located on the left side of the chest, no more than 8 cm (3 in.) to the left of the sternum and over the fourth, fifth, or sixth intercostal space.
33. Rate
34. Rhythm
35. Volume
36. Arterial wall elasticity
37. Presence or absence of bilateral equality.
38. Inhalation-diaphragm contracts, the ribs move upward and outward, and the sternum moves outward, enlarging the thorax. Exhalation-the diaphragm relaxes, the ribs move downward and inward, and the sternum moves inward, decreasing the size of the thorax.
39. Rate
40. Depth
41. Rhythm
42. Qualilty
43. A systolic pressure is the pressure of the blood as a result of contraction of the ventricles. The diastolic pressure occurs when the ventricles are at rest.
44. C
45. A
46. A
47. D
48. C

CHAPTER 19

1. Includes the client's name, address, age, sex, race, marital status, occupation, religious orientation, health care financing, and usual source of medical care.
2. The answer given to the question, "What is troubling you?" or "What brought you to the hospital?"
3. Usual health status
4. Chroniologic story
5. Relevant family history
6. Disability assessment
7. Assists in defining the present problem
8. Biographical data

9. Chief complaint
10. History of present illness
11. Past history
12. Family history of illness
13. Review of systems
14. Life-style
15. Social data
16. Psychologic data
17. Patterns of health care
18. Childhood illness
19. Childhood immunizations
20. Allergies to drugs, animals etc.
21. Accidents and injuries.
22. Hospitalizations for serious illnesses.
23. Medications.
24. Assessing with the senses of sight, smell and hearing. The nurse inspects with the naked eye and with a lighted instrument.
25. The process of listening to body sounds.
26. Body surface is struck to elicit sounds that can be heard or vibrations that can be felt.
27. Using the sense of touch to assess texture, temperature, vibration, masses, distention, pulses and tenderness or pain.
28. Note whether the client is relaxed or tense, has an erect, slouched or bent posture, and has coordinated or uncoordinated movements or tremors.
29. Note whether dress is appropriate to age, life-style, climate, socioeconomic status, culture, or current circumstances.
30. Posture, behavior, and facial expression can reflect distress.
31. Attitude is reflected in appearance, speech, and behavior.
32. Listen for quantity, quality, and organization.
33. D
34. E
35. A
36. G
37. H
38. B
39. F
40. C
41. Myopia
42. Hyperopia
43. Uneven curvature of the cornea
44. Inflammation of the bulbar and palpebral conjunctiva
45. Opacity of the lens
46. Otoscope
47. Weber's test
48. Color, shape, texture and the presence of bony prominences
49. Color, symmetry of contour and texture.
50. Nasal speculum
51. Conduction and sensorineural loss
52. Ask about frequency and severity of headaches, fainting, dizziness, falls and accidents. Is there pain, swelling, stiffness, limited movement or swollen glands?
53. Ask about difficulty of seeing, eye infections, eye pain, excessive tearing, double vision, blurring, sensitivity to light, itching, and spots before eyes. Does the client wear glasses or contacts? When was the last eye examination? Any ear infections, loss of hearing, pain, discharge, or ringing in ears? Does the client wear a hearing aid? Are there frequent colds, nosebleeds, allergies, pain or tenderness in the nose or postnasal drip? Are there sore gums, lumps, white spots in lips or mouth, toothaches, cavities, difficulty swallowing, voice change or hoarseness? Does the client wear dentures? When was the last dental appointment?
54. Is there chest pain, coughing, shortness of breath, wheezing, coughing up blood, or history of lung disease? When was the last chest X-ray? What was the result?
55. Do you have a history of heart disease, palpatations, heart murmur, high blood pressure, anemia, varicose veins, leg swelling or leg ulcers?
56. Liver, gallbladder, duodenum, head of pancreas, right adrenal gland, upper lobe of right kidney, hepatic flexure of colon, sections of the ascending and transverse colon.
57. Left lobe of liver, stomach, spleen, upper lobe of left kidney, pancreas, left adrenal gland, splenic flexure of colon, sections of transverse and descending colon.
58. Lower lobe of right kidney, cecum, appendix, section of ascending colon, right ovary, right fallopian tube, right ureter, right spermatic cord, part of the uterus (if enlarged).
59. Lower lobe of kidney, sigmoid colon, section of descending colon, left ovary, left

fallopian tube, left ureter, left spermatic cord, part of uterus if enlarged.

60. Have you experienced nausea, vomiting, loss of appetite, indigestion, heartburn, bright blood in stools, tarry-black stools, diarrhea, constipation, abdominal pain, excessive gas, hemorrhoids, rectal pain, colostomy, ileostomy?
61. Do you experience muscular pain, swelling, or weakness, joint swelling, soreness, or stiffness, leg cramps or bone defects?
62. Any difficulty walking, unconsciousness, seizures, tremors, paralysis, numbness, tingling, or burning sensations in any body parts? One-sided weakness, speech problems, loss of memory, disorientation, forgetfulness, unclear thinking or changes in emotional state?
63. Frequency, dribbling, urgency, urination at night, difficulty starting stream, blood in urine, incontinence, pain or burning on urination, urinary tract infection, sexually transmitted disease. Age of menstruation, last menstrual period, duration, amount of flow, regularity of cycle? Problems with menstruation, intercourse or vaginal discharge?
64. Frequency, dribbling, urgency, urination at night, difficulty starting stream, blood in urine, incontinence, pain or burning on urination, urinary tract infection, sexually transmitted disease. Penile discharge, swelling, masses or lesions, difficulty in sexual functioning?
65. C
66. D
67. A
68. C
69. B

CHAPTER 20

1. Substances which are lethal to related strains of bacteria
2. The collective vegetation in a given area
3. The ability to produce disease
4. The study of causes
5. T
6. T
7. T
8. T
9. T
10. F
11. T
12. F
13. T
14. F
15. T
16. T
17. T
18. The etiologic agent, or microorganism, the place where the organism naturally resides (reservoir), a portal of exit from the reservoir; a method (mode) of transmission, a portal of entry into a host and the susceptibility of the host.
19. Ensure articles are properly cleaned and disinfected or sterilized before use.
20. Educate clients and family members about appropriate methods to clean, disinfect, and sterilize articles.
21. Dispose of damp, soiled linens.
22. Dispose of feces and urine in appropriate receptacles.
23. Avoid talking, coughing, or sneezing over open wounds or sterile fields.
24. Cover mouth and nose when coughing and sneezing.
25. Place discarded soiled materials in moisture-proof refuse bags.
26. Initiate and implement aseptic precautions for infected clients.
27. Use sterile technique for invasive procedures such as injections and catheterizations.
28. Provide all clients with their own personal care items.
29. Maintain the integrity of the client's skin and mucous membranes.
30. Educate th public about the importance of immunization.
31. Age
32. Heredity
33. Level of stress
34. Nutritional status
35. Immunization status
36. Current medical therapy
37. Preexisting disease processes
38. The time between the entry of the microorganisminto the body and the onset of the symptoms. The length of incubation varies greatly dpending on the microorganism.
39. The time from the onset of nonspecific symptoms until the specific symptoms of the infection appear. Infected persons are

most infectious and most likely to spead the infecting organisms during this stage.
40. During the illness period, specific symptoms develop and become evident. The symptoms are manifested both in the affected body organ or area and in the entire body.
41. The convalescent period extends from the time the symptoms start to abate until the person returns to a normal state of health.
42. Infections that were not present or were incubating at the time the client was admitted to the hospital or other medical facility.
43. They are considered more difficult to prevent and treat, more unpredictable, and more resistant to cure than infections contracted in the community.
44. Localized swelling
45. Localized redness
46. Pain or tenderness with palpation or movement
47. Palpable heat at the infected area
48. Loss of function of the body part affected
49. Fever
50. Increased pulse and respiratory rate if fever is high
51. Lassitude, malaise, and loss of energy
52. Anorexia and possible nausea and vomiting
53. Enlargement and tenderness of lymph nodes that drain the area of infection.
54. Inadequate primary defenses such as broken skin, traumatized tissue, decrease of ciliary action, stasis of body fluids, etc.
55. Verbalizes understanding of individual risk factors
56. Obtains recommended immunizations
57. Wash hands before and after any direct client contact, before any invasive procedure and after contact with any body substance.
58. Place soiled materials in moisture-resistant containers.
59. Dry the skin thoroughly after bathing the client, and apply lotion to roughened or especially dry areas.
60. Ask the client to move, cough, and breathe deeply at least every 2 hours and use aseptic technique when suctioning clients.
61. Use surgical asepsis when changing dressings.
62. Question the client.
63. Observe the client for signs of infection.
64. Review recent laboratory data.
65. D
66. D
67. A
68. C
69. B

Chapter 21

1. Advanced age
2. Hearing loss
3. Previous accidents
4. Walkways and stairways
5. Floors
6. Furniture
7. Bathrooms
8. Kitchen
9. Potential for injury related to impaired physical mobility.
10. The client will begin using a walker within one week.
11. Potential for trauma related to lack of safey precautions with lighted cigarettes.
12. The client will smoke five cigarettes a day in the presence of a family member.
13. Heat to start the fire
14. Combustible material
15. Oxygen to support the fire
16. Hot pot handles
17. Hot bath water
18. Cigarettes
19. Faulty electical wiring
20. Orient to surroundings, and call system.
21. Make sure nonskid bath mats are available in tubs and showers.
22. Encourage her to wear nonskid footwear.
23. Supervise Marjory during the first few days of admission, especially at night.
24. Attach siderails to the bed.
25. Help the client understand that the restraint is a protective device so that he or she does not feel restrained. In most situations a doctor's order is required.
26. Restricts client's movements as little as possible
27. It is least obvious to others
28. It does not interfere with the client's treatment or health problem.
29. It is readily changeable
30. It is safe for the particular client.

31. Tie the ends of the body restraint to the part of the bed that moves when the head is elevated.
32. Apply so that the restraint can be removed quickly in an emergency.
33. Apply so that the client can move as freely as possible.
34. When the restraint is removed, do not leave the client unattended.
35. Provide emotional support verbally and through touch.
36. Keep all cleaning chemicals in a safe place out of the reach of children.
37. Never put chemicals in unmarked food containers.
38. Have all frayed electrical wires and plugs repaired immediately.
39. Teach client never to use electrical appliances around water.
40. Use acoustic tile when remodeling ceilings.
41. Use drapes, carpeting and upholstered furniture to deaden noise.
42. Wear a lead apron when assisting with X-rays or coming in close contact with clients.
43. Wear gloves when handling body waste.
44. A
45. D
46. D
47. B
48. D

CHAPTER 22

1. Promotes cleanliness
2. Provides comfort and relaxation.
3. Improves self-image by improving appearance and eliminating offensive odors
4. Conditions the skin
5. Prepares the client for breakfast or early diagnostic test. The bedridden client is given a urinal or bedpan, the face and hands are washed, and oral care is given.
6. Care is given to clients after breakfast. Includes the provision of a urinal or bedpan, a bath or shower, making the bed, perineal care, back massage, and oral, nail, and hair care.
7. Is provided when clients return from physiotherapy or diagnostic tests. It consists of providing a bedpan or urinal, washing the hands and face, and assisting with oral care.
8. Provided to clients before they retire for the night. It involves providing for elimination needs, washing face and hands, giving oral care, and giving a back massage.
9. Superficial layers of the skin are scraped or rubbed away. The area is reddened and may have localized bleeding or serous weeping. Implications: Wound is prone to infection and should be kept clean and dry. Nurse should not wear rings or jewely to avoid causing abrasions. Lift, do not pull a client across a bed.
10. An inflammatory conditions with papules and pustules. Implications: Keep the skin clean to prevent secondary infection.
11. Redness associated with a variety of conditions: e.g., rashes, exposure to sun, elevated body temperature. Implications: Wash area carefully to remove excess microorganisms. Apply antiseptic spray or lotion to prevent itching, promote healing, and prevent skin breakdown.
12. Alterations in nutritional status
13. Immobility
14. Altered hydration
15. Altered sensation
16. Presence of secretions or excretions on the skin
17. Mechanical devices
18. Altered venous circulation
19. Nursing diagnosis: impaired skin integrity related to immobility. Outcome critieria: changes position every two hours. Intervention: Reposition client and massage bony prominences every two hours.
20. Nursing diagnosis: self-esteem disturbance related to body odor. Outcome criteria: takes a bath or shower daily and uses a deodorant. Intervention: teach client guidelines for daily hygienic self-care.
21. An intact, healthy skin is the body's first line of defense.
22. The degree to which the skin protrects the underlying tissues from injury depends on the general health of the cells, the amount of subcutaneous tissue, and the dryness of the skin.

23. Moisture in contact with the skin for a period of time can result in increased bacterial growth and irritation.
24. Body odors are caused by resident skin bacteria acting on body secretions.
25. Skin sensitivity to irritation and injury varies among individuals and in accordance with their health.
26. Agents used for skin care have selective actions and purposes.
27. D
28. A
29. B
30. F
31. E
32. C
33. Hallux valgus
34. Hammer toe
35. Each foot and toe is inspected for shape, size, and presence of lesions and is palpated to assess areas of tenderness, edema, and cirulatory status. The plantar surface landmarks are the medial longitudinal arch, an apparent heel, and an apparent ball of the foot.
36. Nursing diagnosis: self-care deficit: foot care, related to impaired hand coordination. Outcome criteria: makes routine appointments with a podiatrist. Intervention: teaches client not to use sharp instruments when caring for feet.
37. Nursing diagnosis: pain related to ingrown toenail. Outcome criteria: cuts toenails straight across. Intervention: refer client to podiatrist. Teach proper foot care and measures to prevent future foot problems.
38. The nurse needs a nail cutter or sharp scissors, a nail file, an organge stick, hand lotion or mineral oil and a basin of water. Soak the hand or foot first than cut or file the nail straight across . Avoid trimming or digging into nails at the lateral corners. File to round the corners the gently push back the cuticle.
39. T
40. T
41. T
42. T
43. T
44. F
45. F
46. Bad breath. Implication: client teach or provide regular oral hygine.
47. Cracking of lips. Implication: lubricate lips, use antimicrobial ointment.
48. Gums appear spongy and bleeding. Implication: client teach or provide regular oral hygiene.
49. Lubricate in a sterile, nonirritating wetting solution before inserting. Place a few drops of wetting solution on the lens and spread over both surfaces. To insert hard lenses, ask the client to tilt the head backward, place the lens convex side down on the tip of the dominant index finger, and separate the upper and lower eyelids of the client's eye with the thumb and index finger of your nondominant hand. Place the lens as gently as possible on the cornea directly over the iris and pupil. To insert a soft lens, deep the dominant finger dry for inserting the lens and make sure the lens is not inside out.
50. Use commercially prepared applicators of lemon juice and oil, or a gauze square rolled around the index finger dipped into lemon juice and oil or mouthwash to clean the mucous membranes. Do not use mineral oil because of the danger of aspiration into the lungs. Position the client on the side with the head of the bed lowered and a towel under the chin. Place a curved basin against the chin and lower cheek. Don gloves and brush the teeth. Gently rinse the mouth with a syringe filled with 3 oz. of water or mouthwash.
51. Nursing diagnosis: self-care deficit (oral hygiene) related to low value placed on regular oral care and dental visits. Outcome criteria: brushes and flosses teeth twice a day. Intervention: Teach how to brush and floss teeth correctly.
52. Nursing diagnosis: self-care deficit (contact lens insertion, removal, and cleaning) related to knowledge deficit. Outcome criteria: demonstrates methods of caring for contact lenses. Intervention: teach care for contact lenses properly.
53. Nursing diagnosis: sensory/perceptual alteration: auditory. Outcome criteria: wears a hearing aid throughout the day. Intervention: teach client how to adjust hearing aid for best reception.

54. Nursing diagnosis: ineffective breathing pattern related to excessive secretions in nasopharynx. Outcome criteria: t has patent nares. Intervention: administer anticongestants as ordered by physician.
55. C
56. D
57. B
58. A
59. C

CHAPTER 23

1. Activities directed toward increasing the level of well being and actualizing the potential of individuals, families, and groups; a category separate from primary prevention.
2. Maintaining or improving the general level of health of individuals, families, and groups; part of primary prevention.
3. Individual and community activities to promote healthful lifestyles.
4. Process of enabling people to increase control over and improve their own health; aimed at improving health potential and maintaining health balance.
5. Information dissemination makes use of a variety of media to offer information to the public about the risk of particular life-style choices and personal behaviors. Occurs in the client's home, schools, hospitals or worksites.
6. Apprise individuals of the risk factors that are inherent in their lives in order to motivate them to reduce specific risks and develop positive health habits. Occur in the community, hospitals and worksites.
7. Require the participation of the individual and are geared toward enhancing the quality of life and extending the life span. Occur in community health care institutions and worksites.
8. Serve the needs of individuals spending a great deal of time in the work environment
9. Aimed at the contaminants of human origin that have been introduced into our environment. Occur in a variety of community settings.
10. Model healthy life-style behaviors and attitudes
11. Facilitate client involvement in the assessment, implementation, and evaluation of health goals.
12. Teach clients self-care strategies to enhance fitness, improve nutrition, manage stress, and enhance relationships.
13. Assist individuals, families, and communities to increase their levels of health.
14. Teach clients to be effective health care consumers.
15. Health assessment
16. Formulation of a nursing diagnosis
17. Development of a health promotion/wellness plan
18. Implementation of the plan
19. Evaluation
20. The health history and physical examination provide a means for detecting any existing problems.
21. The physical-fitness assessment includes girth and skinfold measurements, the step test, muscle strength and endurance, and joint flexibility.
22. An assessment and educational tool that indicates a client's risk of disease or injury over the next 10 years by comparing the client's risk with the mortality risk of the corresponding age, sex and racial group.
23. Focuses on the personal life style and habits of the client as they affect health. Categories of lifestyle assessed are nutritional practices, stress management, and other habits.
24. The client's health care beliefs need to be clarified especially those beliefs that determine how they perceive control of their own health care status.
25. Assesses the client's anxiety producing stressors.
26. Nursing diagnoses accepted by NANDA focus on altered health patterns. Wellness oriented nursing diagnoses focus on motivational and wellness behaviors that identify client strengths, recognize self care potential, reinforce healthy lifestyles etc.
27. Identify health care goals
28. Identify possible behavior changes.
29. assign priorities to behavior changes.

30. Make a commitment to change behavior.
31. Identify effective reinforcements and rewards.
32. Determine barriers to change
33. Develop a schedule for implementing the behavior change
34. Individual counseling sessions
35. Telephone counseling
36. Group support
37. Facilitating social support
38. In stage I, unfreezing, the client's motivation to change emerges The nurse must help the client feel safe enough to explore and consider alternatives. In stage II, moving, the client is ready to change and develop new responses. The nurse must take advantage of this stage to elicit new desired client responses. In Stage III, refreezing, the client internalizes behavior changes and stabilizes a new level of functioning. The nurse can assist the client to reinforce new behaviors during this stage.
39. Evaluation takes place on an ongoing basis. The client may decide to continue with the plan, reorder priorities, change strategies, or revise the health promotion contract. It is the collaborative effort between the nurse and the client.
40. D
41. B
42. A
43. D
44. A

CHAPTER 24

1. Growth and development are continuous, orderly sequential processes influenced by maturational, environmental, and genetic factors.
2. All humans follow the same pattern of growth and development.
3. The sequence of each stage is predictable, although the time of onset, the length of the stage, and the effects of each stage vary with the person.
4. Learning can either help or hinder the maturational process, depending on what is learned.
5. Each developmental stage has its own characteristics.
6. Growth is physical change and increase in size and can be measured quantitatively.
7. An increase in the complexity of function and skill progression
8. Sequence of physical changes that are related to genetic influences
9. Sex
10. Physical stature
11. Race
12. Family
13. Religion
14. Culture
15. See Table 24-1, Chapter 24
16. Gesell believes that changes in a child are the result of heredity.
17. Havighurst believes that learning is basic to life and that people continue to learn throughout life; promoted the concept of developmental tasks.
18. D
19. B
20. E
21. C
22. F
23. A
24. Bandura-the entire learning process involves three factors, characteristics of the person, the person's behavior and the environment.
25. Peck -although physical capabilities and functions decrease with old age, mental and social capacities tend to increase in the latter part of life. Gould-seven stages of adult development; transformation is a central theme during adulthood.
26. Piaget -cognitive development is an orderly, sequential process in which a variety of new experiences must exist before intellectual abilities can develop.
27. Kolberg -three levels of moral growth that encompasses six stages. Focuses on the reasons for the making of a decision, not on the morality of the decision itself. Peters-concept of rational morality based on principles. Proposes five facets of moral life. Gilligan-three stages in the process of developing an "ethic of care."
28. Fowler -development of faith is an interactive process between the person and the environment. Westerhoff -four-stage theory of faith development . Proposes that faith is a way of behaving.

29. B
30. B
31. B
32. C
33. D

CHAPTER 25

1. Apgar scoring system
2. Denver Developmental Screening Test
3. Denver Prescreening Developmental Questionnaire
4. Measuring weight, length and height
5. Snellen E chart for vision
6. Assess motor skills, separation behavior, social play behavior, language, nutrition, elimination, and rest/sleep patterns.
7. Measuring weight, height and vital signs
8. Snellen E chart for vision
9. Observing speaking skills for complete sentences
10. Measuring weight, height and vital signs
11. Visual testing
12 Observing psychosocial development
13. The physiological task for infancy is survival which includes breathing, sleeping, sucking, eating, swallowing, digesting, and eliminating. Significant changes occur in weight, length, head growth, vision, and motor development. The central psychosocial task is trust versus mistrust. The infant reaction to stress is crying which is his or her way of communicating distress. Three of the six cognitive stages occur during the first year which include perceptual recognition, memory of certain objects and conceptualization of space and time.
14. Physical development: two-year-olds are usually chubby with short legs and a large head, weighing approximately four times their birth weight. The head circumference is four-fifths of the average adult size. The brain is 70% of its adult size. Toddlers can hold a spoon and are able to run, balance on one foot and ride a tricyle. By three years old, most toddlers are toilet trained. Psychosocial development: the child is in the anal phase according to Freud. Toddlers begin to develop their sense of autonomy and experience separation anxiety and regression. Cognitive development: toddlers develop considerable cognitive and intellectual skills. They understand the sequence of time, have symbolic thought and begin to conceptualize.
15. Physical development: during this time, physical growth slows, but control of the body and coordination increase greatly. Psychosocial development: preschoolers continue to develop self-concept and begin to learn about their feelings and behavior. They have a greater ability to verbalize stress and learn to play and participate with others. Freud describes the Electra and Oedipus complex during this stage. Cognitive development: preschooler's are in the phase of intuitive thought and learn through trial and error. They can count pennies and begin to develop reading skills at this time.
16. Physical development: the school age child gains weight rapidly and appears less thin. The extremities grow more quickly than the trunk. Auditory perception is fully developed. Endocrine function slowly increase between 9 and 13 years and secondary sexual characteristics develop. Psychosocial development: school-age children concentrate on mastering skills for adulthood. They develop significant relationships with others and learn to develop control of their behavior. They tend to use regression, malingering, rationalization and ritualistic behavior to cope with stress.
17. The development task is trust versus mistrust. Little is known about spiritual development at this age. Morally, infants associate right and wrong with pleasure and pain.
18. The developmental task is autonomy versus self-doubt. Toddlers are in the first level of Kohlberg's moral development when children begin to respond to labels of good and bad. Toddlers are aware of religious practices but do not focus on developing spiritual beliefs.
19. The developmental task of the preschooler is initiative versus guilt. Morally they are capable of prosocial (kind) behavior and are in the intuitive-projective stage of spiritual development.

20. The developmental task for the school-age child is industry versus inferiority. Morally the child is motivated to live up to what significant others think of him or her. The child goes through Kohlberg's preconventional and conventional levels of moral development. This child is in the mythical-literal stage and generally believe that God is good.
21 Optimal nutritional status
22. Over the next 2 months, the infant continues to maintain optimal nutritional status.
23 Effective immune response
24. During the next 6 months, the child maintains effective immune response
25. Nutrition intake appropriate for adequate growth
26. Child maintains an appropriate nutritional intake over the next year
27. Positive self concept
28. During the next year, the child continues to maintain a positive self-concept.
29. In the United States the DPTs are given at 2 months, 4 months, 6 months, 15 months and 4 to 6 years. Tetanus toxoid and diptheria toxoid are given between 14 and 16 years. In Canada DPT is given at 2 months, 4 months, 6 months, 18 months, and 4 - 6 years. Diptheria is given between 14 and 16 years.
30. In the United States poliomyelitis immunization is given at 2 months, 4 months, 15 months and between 4 - 6 years. In Canada it is given at 2 months, 4 months, 6 months, 18 months and between 4 - 6 years.
31. In the United States MMR is given at 15 months; at 12 months in Canada.
32. B
33. C
34. B
35. D
36. A

CHAPTER 26

1. Measurement of height and weight
2. Observation of the skin for acne
3. Discussion of personal hygiene needs
4. Measurement of height and weight
5. Assessment of vital signs
6. Assess for hypertension, menstrual problems, sexually transmitted diseases.
7. Measurement of height and weight
8. Assessment of vital signs
9. Assess for visual and hearing changes, cessation of menstruation, and patterns of eating, elimination and exercise.
10. Physical development: growth is accelerated marked by sudden and dramatic physical changes. Adolescent growth spurt in boys usually begins between 12 and 16 years; in girls between 10 and 14 years. Growth is influenced by heredity, nutrition, medical care, illness, physical and emotional environment, family size, and culture. Psychosocial development: the major problems are role confusion and sexual identity. The adolescent needs to establish a self-concept that accepts both personal strengths and weaknesses. Cognitive development: adolescents are in Piaget's formal operations stage of cognitive development. They use logic, organization and consistency in their thinking.
11. Physical development: persons in early twenties are in prime years, physical changes are minimal. Psychosocial development: young adults must make decisions that will affect the rest of their lives, develop relationships and assume social responsibilities. Cognitive development: formal operations period.
12. Physical development: decreasing hormonal production with related physical changes. Psychosocial development: major concern is to provide for others, to be more altruistic. More time for expanding interests. Cognitive development: change very little. Carry out all the strategies in Piaget's phase of formal operations.
13. Developmental task: identity versus role confusion. Spiritual development: synthetic-conventional stage. Moral development: young adolescent is usually at the conventional level but later moves into the postconventional or principled level.
14. Task: intimacy versus isolation. Moral development: postconventional level including principal reasoning. Spiritual

development: individuating-reflective period.
15. Task: generativity versus stagnation. Moral development: conventional level to the postconventional level. Spiritual development: Fowler's paradoxical-consiliadative stage.
16. Satisfying social interactions
17. During the next year, the adolescent continues to maintain health social relationships.
18. Appropriate exercise level
19. During the next 3 months, the young adult continues to maintain exercise status.
20. Positive body image
21. During the next year, the middle-aged adult continues to maintain a positive body image.
22. Vehicle accidents
23. Recreational accidents
24. Vehicle accidents
25. Substance abuse
26. Motor vehicle accidents
27. Work related accidents
28. Assist adolescent deal with skin care especially if acne is a problem.
29. Assist the adolescent deal with nutrition and identify possible eating disorders.
30. Teach breast and testicular self-examination
31. Give industrial safety information.
32. Remind clients to perform monthly breast or testicular self-examination.
33. Encourage influencza and pneumococcal vaccinations if client is in high-risk groups.
34. B
35. B
36. B
37. C
38. C

CHAPTER 27

1. Decreased immune system function, production of saliva, hydrochloric acid and pepsin. Decreased kidney mass and concentrating and diluting ability of the kidney. Decline in male reproduction and cessation of ovum production. Slowing of motor neuron transmission. Decline in autonomic nervous system function. Decrease in sleep stages 3 and 4. Decline in the visual field of the eye and hearing. Decline in sensitivity to taste. Progressive increase in threshold for deep pain. Decreased ability of the body to adapt to stress.
2. May experience mutual withdrawal between the older person and others and maintain their values, habits, and behavior in old age.
3. Continue with Piaget's phase of formal operations. There is minimal change in intellectual capacity of the healthy aging person.
4. Ego integrity versus self-despair according to Erikson. Another theory proposes three developmental tasks; establishing new activities, selecting activities compatible with the physical limitations of old age and making contributions that extend beyond their own lifetime.
5. Most elderly people stay in the conventional level or preconventional level of moral development.
6. Elderly people strive to incorporate views of theology and religious action into thinking. Some enter into the sixth stage of spiritualizing, universalizing.
7. Night driving is hazardous due to lack of accommodation of the eye to light and peripheral vision is diminished. Teach client to turn head before changing lanes. Driving in fog or other hazardous conditions should be avoided.
8. Reduced sensitivity to pain and heat may cause client to be injured by hot packs or heating pads. Encourage client not to use electric pads and to check hot packs carefully.
9. Hyperthermia due to lowered metabolism and thinning subcutaneous tissue. Teach the client to dress warmly with layered clothing, use extra blankets, and eat a balanced diet.
10. B
11. B
12. C
13. B
14. C

CHAPTER 28

1. A family of two parents and their children
2. The relatives of the nuclear families
3 Both parents reside in the home with their children, the mother assuming the nurturing role and the father providing the necessary economic resources.
4. Existing family units who join together to form a new family unit
5. Unrelated people living under one roof
6. Jim, 36-years-old; Luwanda, 35-years-old; Jonella, 12-years-old; Janetta, 14-years-old.
7. Jim, a computer salesman; Luwanda, an accountant; Jonella and Janetta, students
8. Jim's stressors are competition in his job, frequent trips away from home, a sick daughter and a daughter who is disengaged from the family. Luwanda's stressors are Jim's frequent absences, long hours on the job and the same concerns for the children as Jim. Jonella stressors are her illness, being a latchkey kid, being alone too much, parent's absences, and a strained relationship with her sister. Janetta's stressors are school activities, absence of parenting, and strained relationship with her sister.
9. Jim's health problems are hypertension, obesity and peptic ulcers. Luwanda's health problems are obesity, and borderline diabetes. Jonella major health problem is sickle cell anemia. Janetta no major health problems noted.
10. Adaptation - family members are not always aware or supportive of one another's needs.
11. Partnership - father is authoritative and makes most decisions for the family.
12. Growth - the family seems disengaged, each family member is concerned about his or her own interests.
13. Affection - there is a sense of warmth and affection between the parents and the children.
14. Resolve - very little time is spent together as a family. Very little nurturing goes on between the members of the family.
15. Sickle cell anemia
16. Diabetes
17. Hypertension
18. Impaired home maintenance management related to mother returning to work. Outcome criteria: the family will problem-solve issues resulting from mother's employment.
19. Altered family processes related to the illness of the youngest daughter. Outcome criteria : the family members will share in the nurturing and care of the youngest daughter.
20. Families with good communication skills are able to discuss how they feel about the illness and how it affects family functioning. They can enlist a social support network which provides strenth, encouragement, and services to the family during the illness. The crisis of the illness can draw the family together.
21. A
22. A
23. D
24. B
25. D

CHAPTER 29

1. Self-concept is the cognitive component of the self system, whereas self-esteem is the affective component.
2. Body image
3. Role performance
4. Personal identity
5. Self-esteem
6. John's ability to develop trust during early infancy was probably seriously impaired because of neglect and abuse. The development of trust is essential for a healthy self- concept. During later infancy John should have developed autonomy by developing some control over his environment, but was not able to do this in his unhealthy environment.
7. Assertiveness-expressing oneself directly without hurting others. Aggressiveness-attacking others in an angry manner.
8. An individual or group that takes on special importance for the development of self-esteem during a particular life stage.
9. Social role expectations strongly influence people at different stages of life. Expectations differ by age, sex,

socioeconomic status, ethnicity, and career identification.
10. Crises of psychosocial development. The success with which a person copes with achieving developmental tasks determines the development of self-concept.
11. Communication/coping styles. A person's choice of strategies to cope is important in determining whether self-esteem is maintained.
12. Low self-concept
13. She should determine John's perceptions of physical and personal self and observe him for nonverbal cues that reflect his self-perception.
14. Catastrophizing - "I think I have a black cloud over my head."
15. Black and white thinking - complaints of being a failure with crutch walking even though he is making excellent progress.
16. Gina needs to know that John is angry and depressed. His self-esteem and self-worth have been severely damage. He is displacing his anger to her. He is struggling with the affect his amputation will have on his relationship with Gina. His behavior is his way of trying to cope with his loss.
17. The nurse should obtain data related to the number in the family, their ages and residence. Also data about the client's family relationships and satisfaction or dissatisfaction with work roles and social roles.
18. The nurse can help John mobilize the coping skills he used in the past to deal with the crisis in the present.
19. Amputation resulting in refusal to look at stump, and withdrawal .
20. Client goal: accept altered body image. Outcome criteria: looks at the stump by day 3. Verbalizes feeling about body changes by day 2.
21. Establish a trusting relationship. Rationale: mutual trust will foster an atmosphere in which the client will be free to verbalize his feeling. Encourage the client to verbalize his feelings about the amputation. Rationale: feelings must be recognized before they can be dealt with effectively.
22. Depression in response to body image change resulting in anger, isolation and aggressive behavior.
23. Client goal: use assertive rather than aggressive behavior to deal with stress. Outcome criteria: verbalize feelings of powerlessness and anger by day 2. Use assertive behaviors in stressful situations by day 7.
24. Stress positive coping mechanisms and ignore aggressive behaviors client uses. Rationale: positive reinforcement of assertive behavior and extinction of aggressive behaviors will modify inappropriate reactions. Respond to client's aggressive behaviors with positive assertive behaviors. Rationale: observes and models positive assertive behavior displayed by the nurse.
25. C
26. D
27. A
28. A

CHAPTER 30

1. Changes in beliefs, attitudes and behaviors regarding sexuality.
2. The multicultural nature of North American society.
3. The impact of influence groups and movements such as the women's movement, gay liberation, handicapped groups and the "moral majority," on sexuality issues.
4. Sexuality has been with humankind since the beginning of time. The contemporary approaches to sexuality are part of an ever-evolving process.
5. Each ethnic and cultural group has its own traditions that influence the ways they view sexuality. Nurses must be aware that clients from different backgrounds may differ in their approaches to sexuality.
6. People's dealings with sexuality are affected by beliefs and values derived from either religious traditions or some other value system. One function of religion is to provide guidelines for the conduct of sexuality.
7. In today's society there is a great conflict regarding sexuality between religious sects who espouse strict guidelines for sexual

behavior and the more liberal sexual freedom attitude. The nurse must be aware of these differing points of view when caring for clients with different belief systems.

8. Biologic-The newborn infant should have a clearly defined male or female anatomy and physiology. The neurologic, vascular, and other tissues are developed well enugh at birth to allow the sexual organs to respond to stimulation. This can produce penile erection in infant boys and vaginal lubrication in infant girls.
   Psychosocial- Society tend to place infants and small children in role genders. Male and female infants also demonstrate sexual differentiation in motor activity, musculature, attention span, preference for stimuli, and interactions with parent figures.
9. Biologic - Very little change occurs in the anatomic and physiologic components before puberty. As the child progresses in later childhood, body 's begin to develop more muscle and girls develop a slighter structure. It becomes easier to distinguish boys faces from girls faces.
   Psychosocial - Gender identity is a major issue in early childhood. By age 4 or 5 years, the child has a clear sense of being male or female. Gender appropriate behaviors which are reinforced by parents are developed. Gender dysphoria occurs when there is evidence of gender role or gender identity problems.
10. Biologic - Profound changes occur at this time. Refer to Chapter 26 where these changes are discussed in more detail.
    Psychosocial - Include dealing with altered body image, changes in the body's functioning, consolidating gender identity, adjusting gender-role behavior, and learning new social-role behaviors.
11. Biologic - Between 18 and 30 years, the young adult reaches full anatomic and physiologic maturity. These are the prime childbearing and child-rearing years. In the middle years there are changes in hormone levels in both men and women. Psychosocial - Establishment of intimate adult relationships produces change in gender-role expectations. Major components of adulthood sexuality include parenting, role changes and differences in sexual responsiveness.
12. Biologic - The major biologic changes in the older female include continued atrophy of vaginal and breast tissue, decrease and slowing of vaginal lubrication during arousal, decreased vaginal expansion, diminished orgasmic intensity, and a more rapid resolution. Sexual changes in the older male include lowered sperm production, reduction in the size and firmness of the testicles, delay in achieving erection, greater ejaculatory control, less myotonia, reduced orgasmic intensity, more rapid resolution, and longer refractory period.
    Psychosocial - Major issues include adjusting to changing body image, adjusting to changes in family or marital status, retirement, change in body function, and decrease in mobility. Due to societal standards, some elderly clients respond to changes in body image with lowered self-esteem. Widowhood or widowerhood, loss of contact with grown children, and loss of friends have the potential for creating loneliness and depression. Loneliness, lowered self-esteem along with reduced body function and loss of mobility, can lead to social isolation and loss of interaction opportunities.
13. Enlargement of clitoral glans, vaginal lubrication, widening and lengthening of vaginal barrel, separation and flattening of the labia majora, reddening of the labia minora, breast tumescence and enlarged areolae.
14. Retraction of the clitoris, increase in size of the outer one third of the vagina and labia minora, increase in width and depth of the inner two thirds of the vigina, mucoid secretion from the Bartholin's glands.
15. Contraction in the orgasmic platform at 0.8 sec. intervals, contraction of the pelvic floor and uterine muscles.
16. Reversal of vasocongestion
17. Penile erection, tensing, thickening, and elevation of the scrotum, elevation and increase in size of testicles

18. Increase in penile circumference, deepening color, 50% increase in testicular size, mucoid secretions from the bulbourethral glands
19. Rhythmic, expulsive contractions of the penis at 0.8 sec. interval, emission of seminal fluid into prostatic urethra, ejaculation of semen through the penile urethra and expulsion from the urethral meatus.
20. Refractory period and decongestion
21. Information about a client's sexual health should always be obtained during a health history interview. The nurse's professional preparation influences the level of secual health assessment. More indepth data should be collected by a nurse who has advanced or special education in sex education and counseling. Information about sexual problem histories and psychosexual histories should be conducted by professionals who are specialized in sex therapy. Screening for sexual function and dysfunction is conducted by a professional nurse during a health history and can include questions such as, "Do you have any questions about sex or sexual health?" or "What is anything, would you change about your current sexual activity?"
22. Ineffectual or absent role models
23. Altered body structure
24. Lack of knowledge or mininformation about sexuality
25. Physical abuse
26. Value conflict
27. Drugs
28. Depression
29. Puberty and adolescence
30. Drugs
31. D
32. A
33. C
34. F
35. E
36. B
37. Altered sexuality patterns - Verbalizes understanding of sexual anatomy and function.
38. Body image disturbance - Verbalizes concerns about body image, sex role, or desirability as a sexual partner.
39. Avoid sexual contact with persons known to or suspected to have AIDS.
40. Wear a condom during intercourse unless you have a monogamous partner who you know is not infected.
41. Avoid unnecessary bloodtransfusions.
42. Screen all potential blood donors.
43. Recommend that seropositive women delay pregnancy.
44. Provide educational programs on AIDS.
45. The male withdraws the penis from his partner's vagina prior to ejaculation.
46. A covering sheath placed over the penis prior to intercourse.
47. A round rubber cup inserted into the vagina and placed over the cervix.
48. Birth control pills that increase estrogen levels so that ovulation is suppressed.
49. Tying of a woman's fallopian tubes to interrupt tubal continuity.
50. The ligation and cutting of the man's vas deferens on either side of the scrotum.
51. B
52. C
53. B
54. A
55. D

CHAPTER 31

1. Ethnicity refers to individuals who share cultural and social heritage which has been passed from one generation to another, while race refers to the physical differences between groups of people.
2. Material culture consist of objects such as dress, art etc., and the way they are used while nonmaterial culture consists of beliefs, customs, languages, and social institutions.
3. A dominant group is a group within a society which has the authority to function as guardians and sustainers of the value system while a minority group is a group of people who because of their physical or cultural characteristics are singled out from the others in the society in which they live.
4. D
5. F
6. A
7. G

8. B
9. H
10 C
11. J
12. I
13. E

14. Learned
15. Taught
16. Social
17. Adaptive
18. Integrative
19. Ideational
20. Satisfying
21. Most cultures are patriarchal, however the Native American culture is matriarchal. Knowing who the decision maker or dominant person in a family helps the nurse understand the family's decision making process relative to health care.
22. Nurses need to determine if there is a language barrier and enlist the help of an interpreter if necessary.
23. The nurse needs to be aware of the client's comfort index in relation to territoriality.
24. The nurse needs to know whether the client is time oriented to the present or future.
25. How the client views the family may affect health care. For instance, when dealing with a family whose values include the extended family, the nurse needs to consider the needs of the extended family as well as those of the nuclear family.
26. Nursing must be sensitive to the cultural meanings of food and to the foods a client is accustomed to.
27. Americans of African and Mediterranean descent
28. Black and other nonwhite Americans. High incidence in individuals of Taiwan and Japanese descent.
29. Native Americans
30. Breast cancer is more common in white women than black. Skin cancers are less common in black people than in white.
31. Asian and Native Americans convert alcohol into acetaldehyde more rapidly than the general population and therefore experience a rapid onset of and prolonged exposure to high blood acetaldehyde levels which cause alcohol intoxication.
32. Folk medicine, is thought to be more humanistic, takes place in the client's own community or home, is less expensive, more comfortable, and less frightening.

33. Native
34. Asian
35. Black
36. Hispanic
37. Black
38. Native
39. Arab
40. Arab
41. Hispanic
42. Native
43. Asian
44. Arab
45. Black
46. Hispanic
47. Appalachian
48. Black
49. Native
50. Native
51. Appalachian
52. Native
53. Black
54. Hispanic
55. Asian
56. Native
57. Black

58. Because it has its own set of values, language, customs etc.
59. Phase one - honeymoon phase in which people are stimulated by being in a new environment.
60. Phase two - the realization of having to exist in the new enironment accompanied by feelings of frustration and embarrassment.
61. Phase three - seeks new patterns of behavior appropriate to the environment.
62. Phase four - t functions comfortably and effectively.
63. Impaired verbal communications related to language barrier - uses an effective method of communication with verbal and nonverbal cues.
64. Ineffective family coping related to absence of the extended family - employs effective alternative strategies for meeting those needs usually met by the extended family.
65. A
66. D
67. B
68. A
69. D

CHAPTER 32

1. A belief in some higher power, creative force, divine being, or infinite source of energy.

2. A universal- a feature of living, acting, and self-understanding. To believe or be committed to something or someone.
3. An organized system of worship.

4. T
5. F
6. T
7. F
8. T
9. T
10. F
11. F
12. T
13. T
14. T

15. Spiritual distress (pain) related to the death of a loved one - verbalizes acceptance of the loss.
16. Spiritual distress (anger) related to a terminal diagnosis - expresses finding positive meaning in the present situation.
17. Arrange for another Christian scientist to visit client. Do not offer medications or treatments without client's permission.
18. May not drink coffee or tea. Artificial prolongation of life is discouraged.
19. Do pressure to take blood tranfusions. Do not encourage to participate in holidays.
20. Determine whether client follows orthodox Judaism. Arrange for specially prepared meals..
21. Allow family to wash and shroud their dead family member. Do not give pork in the diet. Do not put in a room with a smoking patient since smoking is forbidden.
22. Arrange for priest to visit. Give client opportunity to recieve sacraments, i.e., communication.
23. D
24. D
25. A
26. B
27. C

CHAPTER 33

1. Self-regulating
2. Compensatory
3. Negative feedback systems
4. Several feedback mechanism to correct one physiologic imbalance
5. Adrenocorticotropic hormone (ACTH), which stimulates the adrenal cortex to produce steroids, and thyrotropic hormone (TSH) which stimulates the secretion of thyroxin from the thyroid gland to control the body's rate of metabolism.
6. Antidiuretic hormone (ADH) which controls water reabsorption in the kidney tubules and prevents the body fluids from becoming too concentrated.
7. Epinephrine (adrenaline) and Norepinephrine (noradrenaline) which help the person meet certain emergency situations and support the sympathetic nervous system in the fight-or-flight response.
8. Mineralocorticoids (aldosterone), which induces sodium chloride retention, potassium excretion, and water reabsorption by the kidneys. Glucocorticoids, influence the metabolism of glucose, protein, and fat.
9. Thyroid hormone (TH), which regulates the body's metabolic rate and the processes of growth. Calcitonin which decreases the blood's calcium concentration either by promoting the deposit of calcium into bone or by inhibiting bone breakdown, which would release calcium into the blood.
10. Parathyroid hormone (PTH), which raises plasma calcium levels and lowers plasma phosphate levels.
11. Glucagon which increases blood glucose concentration by stimulating the breakdown of liver glycogen. Insulin which accelerates the movement of sugar, protein, and fats out of the blood and into the tissue cells.
12. Regulates intake of oxygen and exhalation of carbon dioxide.
13. Transports essential elements for all cells.
14. Excretion and absorption of many by-products of metabolism and maintaining acid-base balance.
15. The route for intake of fluids and electrolytes.
16. Stimulus - a life event or set of circumstances causing a disrupted response that increases the individual's vulnerability to illness.
17. Response --the disruption caused by a noxious stimulus or stressor.
18. Transaction - individuals react differently to different situations.

19. The nonspecific response of the body to any kind of demand made on it.
20. Alarm reaction - alerts the body's defenses
21. Stage of resistance - the body's adaptation takes place and attempts to cope with the stressor by limiting it to the smallest area of the body that can deal with it.
22. Stage of exhaustion - the ways that the body used to cope with the stressor during the second stage have been exhausted. The stress effects of the stressor may spread to the entire body.
23. A state of mental uneasiness, apprehension, dread, or foreboding or a feeling of helplessness related to an impending or anticipated unidentified threat to self or significant relationships.
24. An emotional state consisting of a subjective feeling of animosity.
25. Consciously and willfully putting a though or feeling out of mind.
26. Psychologic defensive mechanisms which develop as the personality attempts to defend itself,deal with conflicting impulses, and allay inner tensions.
27. Mild anxiety - a slight arousal state that enhances perception, learning, and productive abilities.
28. Moderate anxiety - increases the client's arousal state to a point where the person expresses feelings of tension, nervousness, or concern.
29. Severe anxiety - perception is furthur decreased. The person focuses on one specific detail of the situation generating the anxiety.
30. Panic, an overpowering, frightening level causing the person to lose control.
31. Physiologic adaptation results in compensatory physical changes such as increase in muscle size following exercise. Psychologic adaptation involves a change in behavior, attitude and coping mechanisms.
32. Diagnosis: anxiety related to the threat of dying. Outcome criteria: verbalizes feelings of anxiety. Interventions: encourage client to participate in plan of care. Give the client the time to express feelings and thoughts.
33. Diagnosis: ineffective individual coping related to work overload. Outcome critiera: identifies personal strengths. Interventions: assist client in developing reasonable self-expectations.
34. Recognize you are stressed
35. Plan daily relaxation
36 Learn to accept failures
37. C
38. D
39. D
40. B
41. A

CHAPTER 34

1. Shock and disbelief - feelings of depression, anger, guilt, sadness, disbelief and denial.
2. Yearning and protest -anger may be directed at God or others whose loved ones are still alive. May fear their own mental deterioration and withdraw from others.
3. Anguish, disorganization and despair - weeping, depression, loss of interest and motivation, inability to make decisions, lack of confidence and purpose.
4. Identification in bereavement - may take on the behavior, personal traits, habits, and ambitions of the deceased. May experience the same symptoms of physical illness.
5. Reorganization and restitution - achieving stability and a sense of reintegration can take a period of time from less than a year to several years.
6. Experiencing the feelings associated with loss before the loved one actually dies. May help family work through grief less painfully.
7. Mutual pretense
8. Mary's answer was appropriate. The nurse should attempt to be as honest as possible when the client asks about death. At this point most clients realize their death is imminent but need to discuss their feelings. Perhaps the client's real question is, "Can I talk about my death?" When a client is dying the nurse should discuss her role with the family and doctor so that she can answer the client's question without fear of repercussion.
9. Ray is angry and depresssed because of his impending death. Mary's youth and

health may remind him of his own weaknesses so he vents his anger on her. Mary should not take his anger personally, since it is a symptom of his own inner turmoil. She can help Ray by reflecting his anger and giving him an opportunity to talk about his feelings. She should stress that his feelings are O.K. and normal for one in his place.

10. Ray is moving through the stages of dying and his emotions may be labile. He is depressed at times and feels hopeless and worthless. His expression may indicate he is angry with God for allowing his illness.
11. Denial: client is unable to accept the impending death.
12. Anger: client expressess anger and hostility over the situation he or she is experiencing.
13. Bargaining: client uses bargaining as a coping mechanism. Bargains with God or a higher being to change the situation he or she is in.
14. Depression: client grieves over loss of life and loved ones.
15. Acceptance: client accepts the changes in his or her life and gradually withdraws from loved ones.
16. Ray's wife is feeling guilty and helpless. Perhaps she feels she has not been a good wife or cared for him adequately. Whatever the reason, she may be projecting her feelings of guilt onto Mary, blaming her for inadequate care. Mary should not take her client's wife's reaction personally. Allowing Ray's wife to vent her feelings and allowing her to participate in Ray's care will help to alleviate her guilty feelings. Referring Ray's wife and daughter to a support group for family members dealing with loss would be helpful.
17. Mary has not worked through her grief over her father's death. Ray's death is reawakening these unresolved feelings of loss for Mary.
18. Ray's family is reacting to the isolation and withdrawal clients often display as they approach death. The nurse needs to explain the client's reaction to the family so that they can offer him needed support at this time.
19. Explain that his family is experiencing difficulty accepting his death. Allow him to verbalize his feelings. Help him to determine how he will handle the problem. Offer assistance by intervening with the family.
20. Hospice is based on holistic concepts that emphasize care rather than cure. Hospice can help Ray and his family by assisting with control and relief of pain and symptoms of the illness and providing social, emotional, and spiritual comfort for he client, family and friends throughout the final stage of illness, at time of death, and during the bereavement period of the survivors.
21. Nursing Diagnosis: fear related to knowledge deficit. Outcome criteria: verbalizes understanding of the seriousness of the illness. Intervention: develop an open and sharing relationship with the client so that he is comfortable asking for information and expressing fears.
22. Nursing Diagnosis: powerlessness related to terminal illness. Outcome criteria: verbalizes feeling of fear, anger etc. Intervention: allow client and family to participate in care.
23. As soon as possible after the death make the body appear natural and as comfortable as possible. Clean the room and remove all equipment and supplies. Clamp or cut all body tubes and tape in place. Place the body in a supine position with the arms at the sides with palms down. Place one pillow under the head and shoulders. Close the eyelids, insert dentures and close mouth. Place absorbent pads under the buttocks. Put on a clean gown and brush and comb the hair. Remove all jewelry except a wedding band which is taped to the finger. Cover the client with clean linens. Turn on a soft light and provide chairs so that family members can view the body.

24. A
25. A
26. D
27. A
28. B

CHAPTER 35

1. Life-style
2. Disability
3. Energy level
4. Age.

5. F	9. A	13. I
6. J	10. E	14. F
7. B	11. C	15. H
8. L	12. D	16. K

17. Pressure - the perpendicular force exerted on the skin by gravity.
18. Friction - the force acting parallel to the skin.
19. Shearing force - a combination of friction and pressure

20. Extension	30. Flexion
21. Flexion	31. Abduction
22. Lateral flexion	32. Circumduction
23. Hyperextension	33. Abduction
24. Circumduction	34. Flexion
25. Circumduction	35. Hyperextension
26. Extension	36. Dorsal flexion
27. Internal rotation	37. Inversion
28. Supination	38. Eversion
29. Abduction	39. Flexion

40. Poorly nourished
41. Decreased sensitivity to pain
42. Existing cardiovascular or pulmonary problems
43. Unconscious
44. Paralysis from either brain or spinal cord injury
45. Reduced level of awareness
46. Malnourished
47. Over 85 years of age
48. Nursing Diagnosis: potential for injury related to joint stiffness. Outcome criteria: demonstrates range-of-motion in all body joints. Intervention: assist client to perform active and passive range-of-motion exercises to maintain mobility of the joints.
49. Nursing Diagnosis: ineffective airway clearance related to shallow breathing. Outcome criteria: takes five deep breaths and coughs every waking hour. Intervention: teach client how to perform diaphramatic breathing to remove trapped stagnant air in the lungs.
50. A
51. D
52. C
53. A
54. C

CHAPTER 36

1. The efficient, coordinated, and safe use of the body to produce motion and maintain balance during activity.
2. It promotes body musculoskeletal functioning, reduces the energy required to move and maintain balance, reduces fatigue and decreases the risk of injury.
3. Body alignment, balance and coordinated body movement
4. A state of equilibrium in which opposing forces counteract each other.
5. An imaginary vertical line drawn through an object's center of gravity.
6. The point at which all of the mass of an object is centered.
7. Labyrinthine sense - sensory organs of the inner ear stimulate postural tonus through impulses that arise when the head is moved.
8. Tonic neck-righting reflexes - movement of the head from side to side affects tonic neck reflexes as well as labyrinthine reflexes.
9. Visual or optic reflexes - visual sensations help the person establish spatial relationships to objects in the environment.
10. Proprioceptor or kinesthetic sense - activated when nerve endings in muscles, tendons, and fascia restimulated by movements of joints.
11. Extensor or antigravity reflexes - extensor muscles which counteract the tendency of the body to flex at the hip and the knees because of its own weight.
12. Plantar reflexes - pressure against the sole of the foot by the ground elicits a reflexive contraction of the extensor mucles of the lower legs.
13. D
14. A
15. C
16. B

17. E
18. Balance is maintained and muscle strain is avoided as long as the line of gravity passes through the base of support.
19. The wider the base of support and the lower the center of gravity, the greater the stability.
20. Objects that are close to the center of gravity are moved with the least effort.
21. Balance is maintained with minimal effort when the base of support is enlarged in the direction in which the movement will occur.
22. The greater the preparatory isometric tensing, before moving the object, the less energy required to move it, and the less the likelihood of musculoskeletal strain and injury.
23. The synchronized use of as many large muscle groups as possible during an activity increases overall strength and prevents muscle fatigue and injury.
24. The closer the line of gravity to the center of the base of support, the greater the stability.
25. The greater the friction against the surface beneath an object, the greater the force required to move the object.
26. Pulling creates less friction than pushing.
27. The heavier an object, the greater the force needed to mover an object.
28. Moving an object along a level surface requires less energy than moving an object up an inclined surface.
29. Continuous muscle exertion can result in muscle strain and injury.
30. Flattened lumbar curve
31. Lardosis
32. Kyphosis
33. Scoliosis
34. A condition in which the bones become brittle and fragile due to calcium depletion.
35. A degenerative disease of the skeletal system causing bony spurs in joints and painful narrowing of the disc spaces of the spine.
36. Reduction in the size in the muscle causing muscle weaknes
37. Nursing Diagnosis: impaired physical mobility related to advanced age. Outcome critera: the client will stand erect when walking. Intervention : encourage client to ambulate in front of a full length mirror to stimulate erect position.
38. Nursing Diagnosis: activity intolerance related to sedentary life-style. Outcome criteria: the client will ambulate one half hour twice a day. Intervention: Teach client to use assistive devices to ambulate.
39. D
40. C
41. B
42. C
43. A

CHAPTER 37

1. Implies calmness, relaxation without emotional stress, and fredom from anxiety.
2. A state of consciousness in which the individual's perception and reaction to the environment are decreased.
3. Biological time clocks which are controlled from within the body and synchronized with environmental factors.

4.	F	9.	T
5.	T	10.	F
6.	F	11.	T
7.	T	12.	T
8.	F	13.	T

14. The stage of very light sleep, characterized by low voltage brain waves. Awakens readily.
15. Stage of light sleep during which the body processes continue to slow down. Heart and respiratory rates and body temperature fall. Lasts 10 - 15 minutes.
16. Heart and respiratory rates continue to slow down. Difficult to arouse.
17. Signals deep sleep. Heart and respiratory rates drop 20 - 30% below those in waking hours. The sleeper is very relaxed, rarely moves, and is difficult to arouse. Physical restoration occurs. Eyes are rolling and some dreaming occurs.
18. 14 - 16 hr/day. 50% REM sleep.
19. 10 - 12 hr/day. 25% REM sleep.
20. 11 hr/day. 20% REM sleep.
21. 10 hr/day. 18.5% REM sleep.
22. 8.5 hr/day. 20% REM sleep.
23. 7 - 9 hr/day. 20 - 25% REM sleep.

24. 7 hr/day. 20% REM sleep.
25. 6 hr/day. 20 - 25% REM sleep.
26. Illness
27. Environment
28. Life-Styl
29. Psychologic Stress
30. Medications
31. Alcohol and Stimulants
32. Diet

33. E
34. G
35. B
36. F
37. A
38. D
39. I
40. C
41. H

42. Causes - alcohol, barbituates, shift work, jet lag, ICU hospitalization, morphine, meperidine. Signs and symptoms - excitability, restlessness, irritability and increased sensitivity to pain, confusion and suspiciousness, and emotional lability.
43. Causes - all of above plus diazepam, flurazepam, hypothyroidism, depression, respiratory distress disorders, sleep apnea and age. Signs and symptoms - withdrawal, apathy, hyporesponsiveness, feeling physically uncomfortable, lack of facial expression, speech deterioration, excessive sleepiness.
44. Usual sleeping pattern, bedtime rituals and use of sleep medications.
45. Time of going to bed, falling asleep, awakening, activities performed before going to bed, and presence of any worries.
46. Observation of the client's facial appearance, unusual behaviors such as irritability, restlessness and physical anomalies causing obstructive sleep apnea.
47. Nursing diagnosis: sleep pattern disturbance: insomnia related to change in sleep environment. Outcome criteria: the client will fall asleep within 30 minutes of going to bed. Intervention: close door to client's room, defer telephone calls with client's permission, plan treatments and medications before bedtime so that noise and hospital routines are minimized.
48. Sleep pattern disturbance: hypersomnia related to multiple stresses. Outcome criteria: verbalizes less depression and stress. Intervention: encourage client to verbalize feelings. Assist to problem solve and find alternative ways of dealing with problems. Reinforce healthy coping mechanisms the client used in the past.
49. B
50. C
51. B
52. A
53. D

CHAPTER 38

1. Is intense and generally of short duration
2. Develops more slowly and lasts much longer than zcute pain
3. Is resistant to cure or relief
4. Is actual pain felt in a body part that is no longer present such as an amputated foot
5. Is perceived at the source and extends to surrounding or nearby tissues
6. Arises from the skin, mucles or joints. It may be superficial or deep
7. Results from stimulation of pain receptors in the abdominal cavity and thorax
8. Travels along the same pathways as viseral pain which is perceived as somatic pain
9. Reception, transmission/perception and modulation
10. The amount of pain stimulation a person requires before feeling pain
11. Pain sensation
12. Chemical regulators that may modify pain
13. The maximum amount and duration of pain that an individual is willing to endure
14. The autonomic nervous system and behavioral responses to pain
15. Bradykinin an amino acid chain that causes powerful vasodilation and increased capillary permeability, constricts smooth muscle, and stimulate pain receptors.
16. Specificity theory - assumes that pain travels from a specific nociceptor to a pain center in the brain.
17. Pattern theory - includes the peripheral pattern theory , the central summation theory and the sensory interaction theory.
18. Gate-control theory - peripheral nerve fibers carrying pain to the spinal cord can have their input modified at the spinal

cord level before transmission to the brain. Synapses in the dorsal horns act as gates that close to keep impulses from reaching the brain or open to permit impulses to ascent to the brain.

19. Parallel processing model - the physiologic or neurologic deciphering of the pain sensation and the cognitive-emotional properties of pain occur along different nerve fibers. Pain is processed at three levels.
20. Nursing diagnosis: chronic pain related to fatigue. Outcome criteria: modifies activities according to limitations. Intervention: Plans daily activities with client aimed at conserving energy.
21. Nursing diagnosis: fear related to intractable pain. Outcome criteria: verbalizes use of relaxation techniques to control pain. Intervention: teach how to use relaxation techniques.
22. B
23. A
24. C
25. D
26. C

CHAPTER 39

1. Nutrition
2. Nutrients
3. Nutritive value
4. Calorie
5. A large calorie
6. The basal metabolic rate (BMR)
7. Carbon, hydrogen and oxygen
8. Monosaccharides and Diasaccharides
9. Natural
10. Refined or processed
11. Fiber.
12. Catalysts
13. Amino acids
14. Essential amino acids
15. Meats, poultry, fish, dairy products and eggs
16. Positive nitrogen balance
17. Lipids
18. Heart disease
19. Fat-soluble vitamins
20. Calcium and phosphorus
21. Microminerals
22. Gertrude's nutritional status is poor as evidenced by her subjective and objective signs and symptoms. Subjective data- fatigue, weakness, and blurred vision . Objective data- frailty, confusion, dehydration, reduced elasticity, edentulous, pale, falling.
23. Some physiologic reasons for malnutrition in the elderly are slowed down gastrointestinal processes, lack of teeth, poor dental status, poorly fitting dentures, sedentary life-style, chronic illness.
24. Depression and loneliness, inability to shop or prepare foods, confusion or forgetfulness, fixed or inadequate monthly income
25. Gertrude's nutritonal problems will not be solved by ths diet. Her diet is deficient in foods from the milk and vegetable groups. She is not taking enough fluid. Her diet is too high in fats and deficient in protein and fiber.
    Lack of protein creates a negative nitrogen balance, which impedes wound healing. Vitamin C is needed for good wound healing and tissue repair.
26. Nursing diagnosis: altered nutrition less than body requirements related to advanced age resulting in loss of weight, dizziness, fatique, weakness and falling. Outcome criteria: agrees to eat meals prepared for the elderly in the community such as "Meals -On-Wheels." Intervention: make a referral through the social service department or community agency to arrange to have prepared meals delivered to client's home.
27. Nursing diagnosis: potential for injury related to malnourishment resulting in weakness, fatique and falling. Outcome criteria: eats at least 1500 calories from the four basic food groups daily. Intervention: teach client the amount of servings she needs from each of the basic food groups to meet her daily body requirements.
28. D
29. D
30. A
31. D
32. C

CHAPTER 40

1. Fluid found within the cells of the body and constitutes two-thirds to three-quarters of the total body fluid
2. Fluid found outside of the cells and is composed of intravascular and interstitial plasma
3. Fluid found within the vascular system
4. Fluid that surrounds the cells including lymph
5. The product of a gland
6. Waste produced by the cells of the body
7. 57%, 40
8. Highest
9. Decreases
10. Greater

11. T
12. T
13. F
14. T
15. F
16. T
17. T
18. T
19. T

20. Urine
21. Insensible loss through the skin as perspiration and through the lungs as water vapor.
22. Noticeable loss through the skin as sweat
23. Loss through the intestines in feces
24. 135-145 mEq/L. Deficit: monitor Intake and output. Assess for anorexia, nausea, vomiting. Encourage foods and fluids containing sodium. Excess: monitor intake and output. Check for symptoms. Encourage intake of fluids. Teach diet low in sodium.
25. 3.6-5.0 mEq/L. Deficit: monitor cardiac changes, ie weak , irregular pulse. Administer potassium as ordered. Teach about food sources high in potassium. Monitor intake and output. Excess: monitor cardiac function for irregular pulse rate and bradycardia. Restrict potassium in diet. Monitor serum potassium.
26. 4.3-5.3 mEq/L. Deficit: administer calcium supplement as needed. Advise to increase or decrease regular dietary calcium. Initiate seizure precautions. Monitor serum calcium. Excess: assess client for signs of hypocalcemia. Monitor serum calcium. Assess client for signs of hypercalcemia. Inspect urine for calculi.
27. 98-108 mEq/L. Deficit: usually associated with hyponatremia. Monitor serum potassium. Excess: usually associated with hyper natremia. Monitor serum potassium.
28. 1.5-2.5 mEq/L. Monitor breathing. Employ safety precautions if confusion or seizures are anticipated. Assist with magnesium replacedment as ordered. Excess: monitor vital signs and level of consciousness.
29. 1.2-3.0 mEq/L. Deficit: monitor serum phosphate levels. Assess neurologic signs and orientation. Administer IV phosphate as ordered. Teach client about food sources high in phosphorus. Excess: Administer magnesium or calcium or antacids as ordered. Assist with diet low in phosphorus.
30. The kidneys excrete hydrogen ions and form bicarbonate ions in specific amounts as indicated by the pH of the blood. When the plasma pH drops, hydrogen ions are excreted, and bicarbonate ions are formed and retained. When the plasma pH rises, hydrogen ions are retained in the body, and bicarbonate ions are excreted.
31. Nursing diagnosis: fluid volume deficit, actual related to elevated temperature. Outcome criteria: temperature will return to normal limits within 8 hours. Intervention: increase oral fluid intake. Administer antipyretic drugs as ordered.
32. Nursing diagnosis: fluid volume excess related to excess sodium intake. Outcome criteria: five pound of weight reduction in one week. Intervention: weigh client daily. Observe fluid intake and output.
33. A
34. B
35. A
36. D
37. B

CHAPTER 41

1. The process of gaseous exchange between the individual and the environment.
2. Pulmonary ventilation - the inflow and outflow of air between the atmosphere and the alveoli of the lungs.

3. Diffusion of gases between the alveoli and pulmonary capillaries.
4. Transport of oxygen and carbon dioxide via the blood to and from the tissue cells.
5. Tidal volume
6. Total lung capacity
7. Intrapulmonic pressure
8. Intrapleural pressure
9. Lung compliance
10. Lung recoil
11. Surfactant
12. Diffusion
13. Hemoglobin
14. Cardiac output
15. Erythrocytes
16. Hematocrit
17. The respiratory center
18. The dorsal respiratory group
19. Environment
20. Exercise
21. Emotions
22. Lifestyle
23. Health status
24. Narcotics
25. Increased pulse rate or increased rate and depth of respirations
26. Slight increase in systolic blood pressure
    Possible cyanosis
27. Decreased pulse rate
    Decreased systolic blood pressure
28. Cough
    Hemoptysis

29. C
30. I
31. H
32. K
33. A
34. G
35. F
36. B
37. E
38. L
39. J
40. M
41. D

42. Low pitched snoring during inhalation
43. Extreme inspiratory effort that produces no chest movement
44. Marked sternal and intercostal retractions
45. Adequate hydration maintains the moisture of the respiratory mucous membranes. If the client becomes dehydrated the mucous membranes secrete mucus that is thick and tenacious. The mucous membranes then become irritated and prone to infection.
46. They measure the flow of air inhaled through the mouthpiece. They therefore offer an incentive to improve inhalation.
47. A lung inflation technique for delivering air or oxygen into the lungs at positive pressure during inspiration and automatic releasing of the pressure when the predetermined positive pressure level is reached in the air passages. May be more effective in expanding the lungs, moving secretions, promoting coughing, and delivering aerosol medications into the deeper, smaller air passages than the incentive spirometers.
48. Nursing diagnosis: ineffective airway clearance related to dehydration. Outcome criteria: increases fluid intake to 2400 ml per day. Intervention: teach the importance of an adequate fluid volumn intake. Keep fluids available at bedside at all times. Encourage fluid intake.
49. Nursing diagnosis: ineffective breathing pattern related to decreased lung expansion. Outcome criteria: is free of cyanosis. Intervention: encourages deep breathes every hour. Teachehow to use the incentive spirometer.
50. A
51. D
52. C
53. A
54. D

CHAPTER 42

1. Wave-like movements produced by the circular and logitudinal muscle fibers of the intestinal walls which propels the intestinal contents forward
2. Waste products that leave the stomach, pass through the small intestine and the ileocecal valve
3. Movement of the chyme back and forth within the haustral pouches of the large intestine
4. Mouth
5. Epiglottis
6. Duodenum
7. Hepatic flexure
8. Transverse colon
9. Ascending colon
10. Cecum

11. Appendix
12. Anus
13. Esophagus
14. Stomach
15. Splenic flexure
16. Descending colon
17. Sigmoid
18. Rectum

19. Age and development
20. Diet
21. Fluid
22. Activity
23. Psychologic factors
24. Life-style
25. Medications
26. Diagnostic procedures
27. Anesthesia and surgery
28. Pathologic conditions
29. Irritants
30. Pain
31. Color
32. Consistency
33. Shape
34. Amount
35. Odor
36. Abnormal constituents
37. Causes: absence of bile pigment/Liver or gallbladder disease. Implication: observe for other gastrointestinal symptoms. Look for signs of bleeding such as petechiae, bleeding gums, hematuria.
38. Causes: drug (e.g., iron), bleeding from uupper gastrointestinal tract, diet high in red meat and dark green vegetables. Implications: observe vital signs frequently. Report physician immediately.
39. Causes: bleeding from lower gastrointestinal tract, some foods (e.g., beets). Implications: observe vital signs frequently. Report to physician immediately.
40. Causes: excessive bile in the GI tract caused by infection. Undigested vegetables. Implications: collect specimen for culture and sensitivity. Use handwashing and stool precautions.
41. Cause: bacterial infection. Implication: same as Number 46.
42. Nursing diagnosis: constipation related to immobility. Outcome criteria: incorporates daily exercise into lifestyle. Intervention: discusses the importance of exercise with the client. Assists in plan to include daily exercise.
43. Nursing diagnosis: diarrhea related to food allergies. Outcome criteria: avoids foods known to be irritating to the bowel. Relates the onset of diarrhea to the ingestion of a new food. Intervention: teachhow to determine food sensitivity.
44. C
45. C
46. A
47. B
48. B

CHAPTER 43

1. The process of emptying the bladder
2. Developmental variables
3. Psychosocial factors
4. Fluid and food intake
5. Medications
6. Muscle tone and activity
7. Pathological conditions
8. Surgical and diagnostic procedures
9. Polyuria-increased production in urine caused by some forms of diabetes, increased intake in fluid, and caffiene.
10. Anuria-absence of urine production due to kidney disease or trauma.
11. Nocturia-urination during the night due to bladder infection or other urinary elimination problems.
12. Frequency-increased voiding times during the day and night due to cystitis or stress.
13. Enuresis-bedwetting due to psychological stress or kidney/bladder disorder.
14. Incontinence-inability to hold urine due to neurological disease or physical truama.
15. Ureters are diverted to the abdominal wall or flank and a ureteral stoma is formed.
16. A segment of the ileum is removed and the intestinal ends are reattached. A pouch and stoma are created.
17. A tube is formed from part of the bladder wall. A stoma is formed from at one of the tube.
18. Ureters are implanted into the signmoid colon.
19. Developmental variables
20. Patterns of elimination
21. Alterations in elimination
22. Physical assessment
23. Urine

24. Diagnostic tests
25. Past illnesses or surgery
26. 1200-1500 ml; unders 1200 ml; kidney failure, reduced intake.
27. Not present; present; diabetes
28. 4.5-8; over 8, under 4.5; urinary tract infections, uncontrolled diabetes
29. Not present; present; cystitis.
30. Diagnosis: stress incontinence related to weak bladder muscles. Outcome criteria:performs Kegel's exercises four times a day. Intervention: teaches client how to perform exercises.
31. Diagnosis: disturbance of self-concept related to incontinence. Outcome criteria: verbalizes interest in urine diversion surgery. Intervention: provides information regarding client's options.
32. A
33. A
34. B
35. C
36. C

CHAPTER 44

1. T
2. F
3. T
4. F
5. F
6. T
7. Development
8. Culture
9. Stress
10. Medications
11. Illness
12. Life-style
13. Results when the level of sensory input is too low to function.
14. Results when more sensory stimulation is experienced in a given period than one can tolerate.
15. Impaired functioning of a sensory or perceptual process
16. Changes in attention span such as decreased concentration, increased distractibility, restlessness, and daydreaming.
17. Changes in thought processes such as confusion about time, place, or person, disordered sequencing of time or events, difficulty in remembering, difficulty in grasping ideas etc.
18. Emotional lability, such as rapid mood swings, or irritability.
19. Changes in usual routines, such as altered sleeping pattern.
20. Nursing diagnosis: altered thought processes related to sensory deprivation. Outcome criteria: maintains orientation to time, place, and person. Intervention: use orientation stimuli in the environment such as calendars, and television.
21. Nursing diagnosis: potential for injury related to visual impairment. Outcome criteria: learns to use a cane and other methods for ambulating in the environment. Intervention: refer to school for the blind if appropriate. Teach client how to use a cane for safety.
22. D
23. B
24. C
25. A
26. D

CHAPTER 45

1. Medication
2. Prescription
3. Official name
4. Trademark or brand name
5. Pharmacology
6. Pharmacist
7. A book containing a list of products used in medicine, with descriptions of the product, chemical tests and formulas
8. A book that containing drugs designated as official by the Federal Food, Drug, and Cosmetic Act
9. Is a collection of formulas and prescriptions
10. Nurses need to know how nursing practice acts in their areas define and limit their functions and also be able to recognize the limits of their own knowledge and skill.
11. The inappropriate intake of a substance, either continually or periodically
12. A person's reliance on or need to take a drug or substance
13. Due to biochemical changes in body tissues.

14. Emotional reliance on a drug to maintain a sense of well-being, accompanied by feelings of need or cravings for that drug

15. D
16. G
17. E
18. H
19. B
20. A
21. I
22. J
23. F
24. C
25. 50%
26. 25%
27. 12.5%
28. 6.25%

29. Oral intake of medication by mouth. It is the most common, least expensive and most convenient and safe route.
30. A drug is placed under the tongue and is absorbed into the blood vessels on the underside of the tongue in a relatively short time.
31. A drug is held in the mouth against the mucous membranes of the cheek until the drug dissolves.
32. Admininstration of a drug into the subcutaneous tissue, just below the skin.
33. Administration of a drug into the muscle
34. Administration of a drug under the epidermis.
35. Administration of a drug into a vein
36. Full name of the client
37. Date the order is written
38. Name of the drug to be administered
39. Dosage of the drug
40. Method of administration
41. Signature of the physician or nurse practioner
42. Right drug
43. Right dose
44. Right time
45. Right route
46. Right client
47. A
48. C
49. A
50. A
51. D

CHAPTER 46

1. An injured body area
2. A wound that occurs during therapy
3. A wound that occurs as a accident
4. A wound that is made as a result of a sharp instrument such as a scapel or knife
5. A closed wound that is the result of a blow from a blunt instrument
6. An open wound that results from friction
7. An open wound made by a sharp instrument that penetrates the skin and underlying tissues
8. Tissues are torn apart producing irregular edges
9. Vasculature
10. Immune status
11. Nutrition
12. Preoperative stay
13. Preoperative preparation
14. Intraoperative elements
15. Increased pulse rate, increased respiratory rate, lowered blood pressure, restlessness, thirst and cold, clammy skin
16. Redness, swelling, pin, induration, fever, increased leukocyte count
17. Unexplained fever and tachycardia, unusual wound pain and prolonged paralytic ileus
18. Visual inspection, palpation, and the sense of smell
19. Nursing diagnosis: potential for infection related to impaired skin integrity. Outcome criteria: maintain intact skin around drainage site. Intervention: keep dressings clean and dry. Cleanse skin of all exudate when changing dressings.
20. Nursing diagnosis: body image disturbance related to altered body function. Outcome criteria: verbalizes feelings about change in body function. Intervention: Encourages client to verbalize feelings. Refers to appropriate rehabilitation programs.
21. C
22. A
23. C
24. B
25. D

CHAPTER 47

1. The time before, during, and after an operation
2. Begins when the decision for surgical intervention is made, and ends when the client is transferred to the operating room bed.
3. Begins when the client is transferred to the operating room bed, and terminates

when the client is admitted to the postanesthetic area.

4. Begins with admission to the Post Anesthetic Care Unit

5. C
6. H
7. E
8. B
9. I
10. D
11. J
12. G
13. A
14. K
15. F

16. Age: The very young and elderly are more prone to surgical risk.
17. Nutritional status: obesity and malnutrition due to protein, iron, and vitamin deficiencies can increase surgical risk.
18. Fluid and electrolyte status: dehydration and hypovolemia predispose a client to problems during surgery.
19. General health: surgery is least risky when the general health is good.
20. Medications: anticoagulants, tranquilizers, heroin and other depressants, antibiotics, and diuretics may adversely affect surgery.
21. Mental health and attitude: extreme anxiety can impact on surgical outcomes.
22. Nursing diagnosis: fear related to risk of death. Outcome criteria: expresses feeling about the surgery and its expected outcome. Intervention: encourage client to ventilate feelings. Set up visit by clergy with client's permission.
23. Nursing diagnosis: sleep pattern disturbance related to psychologic stress. Outcome criteria: verbalizes feelings about impending surgery. Sleeps uninterrupted for at least six hours. Interventions: Carries out measures to ensure client's comfort. Plan nursing activities so that client is undisturbed after bedtime. Reduces external stimuli in the environment.
24. Scrub nurse: handsthe surgeon sterile instruments, counts sponsges, needles, and instruments and disposesof used instruments.
25. Circulating nurse: helps to position and drape the client, opens sterile packages, sends biopsy specimens to the laboratory, adjusts operating room lights and obtains additional supplies and equipment.
26. Respiratory function, cardiovascular function, fluid and electrolyte balance, dressing, tubes, and drains, neurologic status, pain and safety.
27. Nursing diagnosis: impaired physical mobility secondary to incisional pain. Outcome criteria: requests pain medication when needed. Reports no severe pain. Intervention: assesses for pain and administered prescribed medication according to physician's order.
28. Nursing diagnosis: potential ineffective airway clearance related to history of smoking. Outcome criteria: performs deep-breathing exercises and voluntary coughing every two hours when awake. Intervention: teaches client how to perform coughing and deep breathing exercises.
29. Collapse of the alveoli, with retained secretions. Signs: marked dyspnea, cyanosis, pleural pain, prostration, tachycardia, increased respiratory rate, fever, productive cough, ausculatory crackling sounds. Interventions: deep-breathing and coughing exercises, turning, early ambulation, adequate fluid intake.
30. Bleeding internally or externally. Signs: rapid weak pulse, increasing respiratory rate, restlessness, lowered blood pressure, cold clammy skin, thirst, pallor, reduced urine output. Intervention: early recognition of signs.
31. Inflammation of bladder. Signs: burning sensation when voiding, urgency, cloudy urine, lower abdominal pain: Adequate fluid intake, early ambulation, good perineal hygiene.
32. Separation of a suture line before the incision heals. Signs: increased incision drainage, tissues underlying skin become visible along parts of the incision. Intervention: adequate nutrition, appropriate incisional support and avoidance of strain.
33. D
34. B
35. A
36. D
37. B

CHAPTER 48

1. Before the test the nurse should explain what the test is, where it will take place, who will do it, how long it will last, and when the results will be available. The informed consent has to be signed by the client after the physician has explained the test. The nurse assembles the equipment and has it available for the doctor. Baseline data should be obtained for assessment during and following the procedure. Any undue anxiety on the the client's part should be reported to the physician. During the test the nurse continues to assess the client for an untoward signs and provide emotional support. Any distress needs to be reported immediately to the physician conducting the procedure. After the test the nurse should label and send any specimen obtained to the laboratory. The client's vital signs should be assessed periodically at regular intervals until stable and be placed in a comfortable or prescribed position. The time of the treatment or test, who performed it, the specimen obtained, and the client's response to the test should be recorded.
2. A sterile procedure using a bronchoscope to examine and biopsy the bronchi. During the procedure the nurse assists the doctor as needed. Vital signs are monitored and the client is given support as needed. After the procedure the vital signs are monitored every 30 minutes until stable, fluids and foods are withheld until the gag reflex is restored and the client is conscious, and the client is positioned in the lateral position until conscious.
3. A visual examination of the interior of urinary bladder with a cystoscope. Medications may be instilled and biopsies may be taken. During the procedure the client is given support, vital signs are monitored, and the physician is assisted as needed. After the procedure the vital signs, urination and urine are monitored, the unconscious client is positioned, and the fluids intake is increased.
4. Examination of the interior of the rectum with a proctoscope. During the examination support the client in the knee-chest position, monitor vital signs and label and send specimens to the pathology laboratory. After the examination monitor vital signs, inspect the next few stools for blood, allow the client to rest.
5. The client swallows barium, and x-ray films are taken of its course through the pharynx, esophagus, stomach, duodenum, and large intestines. The nurse encourages fluids and activity to prevent constipation, a laxative may have to be administered, and the stool is observed for the passage of the barium.
6. X-ray films are taken of the bile ducts after dye has been administred intraveneously. The nurse assesses for allergy to the dye and the IV site is observed for bleeding and tenderness.
7. X-ray films are taken of the gallbladder after a contrast dye has been given orally. After the procedure the nurse provides a rest period for the client, the normal diet is restored, and the client is assessed for allergy to the contrast dye.
8. A contrast material is injected into the subarachnoid space, and x-ray films are taken of the spinal cord, nerve roots, and vertebrae. After the procedure the client is positioned flat in bed for 24 hours to minimize headache and/or nausea, vital signs are monitored as well as the urinary output.
9. B
10. B
11. A
12. D
13. B